Modern Frontend Development with Node.js

A compendium for modern JavaScript web development within the Node.js ecosystem

Florian Rappl

BIRMINGHAM—MUMBAI

Modern Frontend Development with Node.js

Group Product Manager: Pavan Ramchandani
Publishing Product Manager: Bhavya Rao
Senior Editor: Divya Anne Selvaraj
Technical Editor: Joseph Aloocaran
Copy Editor: Safis Editing
Project Coordinator: Manthan Patel
Proofreader: Safis Editing
Indexer: Hemangini Bari
Production Designer: Joshua Misquitta
Marketing Coordinator: Anamika Singh

First published: November 2022
Production reference: 1171122

Published by Packt Publishing Ltd.
Livery Place
35 Livery Street
Birmingham
B3 2PB, UK.

ISBN 978-1-80461-829-5
www.packt.com

Dedicated to every web developer out there. Always keep learning, keep innovating, and keep sharing. Thanks for all your efforts and hard work!

– Florian Rappl

Contributors

About the author

Florian Rappl is a solution architect working on distributed web applications for digital transformation and IoT projects. His main interest lies in the implementation of micro frontends and their impact on teams and business models. In this area, he has led several teams, realizing many successful projects over the last few years.

As the lead architect, he has helped to create outstanding web applications for many industry-leading companies. He regularly gives lectures on software design patterns and web development. Florian has won multiple prizes for his work over the years and is recognized as a Microsoft MVP for development technologies.

He started his career in software engineering before studying physics and helping to build an energy-efficient supercomputer. Florian currently lives in Munich, Germany, with his wife and their two daughters.

About the reviewers

Israel Antonio Rosales Laguan is an experienced full stack software engineer using JavaScript, React, and Node.js with a focus on process improvement, developer ergonomics, systems integration, and pipeline automation. He also has strong experience in international SCRUM teams and mentoring others, working in Equinox, OnDeck, and Lazard, among others. Other expertise includes OWASP compliance, GraphQL, CI/CD with Docker, and advanced CSS.

Abhishek Kumar Maurya is a senior software engineer with 12+ years of industry experience. He currently works at Rippling, but has worked at various organizations such as Oracle India Pvt Ltd and founded his own start-up. He graduated from Banaras Hindu University, Varanasi, and post-graduated from NIT Trichy. He has worked on both the backend and frontend and is currently focusing on the frontend with Node.js and React.

Table of Contents

Part 2: Tooling

4

Using Different Flavors of JavaScript   51

5

Enhancing Code Quality with Linters and Formatters   65

6

Building Web Apps with Bundlers   79

7

Improving Reliability with Testing Tools   105

Part 3: Advanced Topics

8

9

10

11

Using Alternative Runtimes   169

Preface

This book covers everything necessary to make you use the power of Node.js, its concepts, and its ecosystem to the fullest. This includes everything you need to know about module systems, packages, helper libraries, CLI tools, WebAssembly, and a range of available tools such as bundlers (Webpack (v5), Parcel (v2), Vite, and esbuild), test runners (AVA, Jest, and Mocha), transpilers (Babel and TypeScript), and many more tools (Flow, Prettier, eslint, and Stylelint) are also covered.

Who this book is for

This book is for junior and intermediate-level frontend web developers who are looking to leverage the Node.js ecosystem for building frontend solutions. The book requires beginner-level knowledge of JavaScript, HTML, and CSS. Prior experience in using a standard shell (sh) will be beneficial.

What this book covers

Chapter 1, *Learning about the Internals of Node.js*, describes the inner workings of Node.js, its principles, and basic ideas. This chapter also makes you familiar with the essential Node.js command-line tooling.

Chapter 2, *Dividing Code into Modules and Packages*, introduces different module formats, their advantages and disadvantages, and their support within Node.js. The chapter also introduces the important `package.json` file for defining Node.js packages.

Chapter 3, *Choosing a Package Manager*, describes and compares the different established command-line utilities for installing and managing third-party dependencies in your Node.js packages.

Chapter 4, *Using Different Flavors of JavaScript*, covers the main concepts and ideas for using different flavors of JavaScript with Node.js. These flavors include Flow and TypeScript, but also more recent specifications of the ECMAScript standard than those supported by the currently available version of Node.js.

Chapter 5, *Enhancing Code Quality with Linters and Formatters*, covers the available utilities for improving the code quality of JavaScript projects. This chapter has information on how to install these code quality helpers, configure them, and have them integrated into standard workflows and development processes.

Chapter 6, *Building Web Apps with Bundlers*, discusses everything you need to know about dedicated web build tools known as bundlers. In this chapter, you will learn how state-of-the-art web projects are compiled from source code to artifacts that can be published on a server. The covered bundlers include Webpack, esbuild, Parcel, and Vite.

Chapter 7, Improving Reliability with Testing Tools, covers everything you need to know about testing with Node.js – from tools for running unit tests to full end-to-end test runners. In particular, the chapter includes elementary knowledge about Jest, Mocha, AVA, Playwright, and Cypress.

Chapter 8, Publishing npm Packages, contains useful information to publish and consume packages from the official npm registry or a custom private registry such as Verdaccio. The chapter also covers the creation and publishing of CLI tools with Node.js, as well as information about writing isomorphic libraries.

Chapter 9, Structuring Code in Monorepos, covers general strategies for the development of multiple dependent packages with Node.js. In particular, it goes into the details of working on multiple packages within a single repository known as a monorepo. Possible tools, such as Nx, Lerna, or Turbo, are introduced in combination with npm, Yarn, and pnpm workspaces.

Chapter 10, Integrating Native Code with WebAssembly, discusses the possibility of running native code compiled to WebAssembly. The chapter guides you through creating your first WebAssembly module, as well as running the created module in the browser and in Node.js.

Chapter 11, Using Alternative Runtimes, offers a detailed view of two alternatives to Node.js: Deno and Bun. Both are evaluated in terms of compatibility, security, performance, and stability.

To get the most out of this book

All the examples in the book have been created with simplicity in mind. They all work similarly and only require knowledge in core frontend technologies such as JavaScript with HTML and CSS. Additionally, some basic knowledge in using a terminal is necessary in order to follow all examples. The tooling to make the code run is discussed throughout the book. As such, if you know how to work with JavaScript, and follow the book explaining how to use Node.js with npm, you'll have no problems running the examples presented in the book.

Software/hardware covered in the book	Operating system requirements
Node.js 14 or higher	Windows, macOS, or Linux
npm 6 or higher	
ECMAScript 2015 (6)	

In *Chapter 11*, you'll also run Deno and Bun. The chapter itself contains installation instructions.

If you are using the digital version of this book, we advise you to type the code yourself or access the code from the book's GitHub repository (a link is available in the next section). Doing so will help you avoid any potential errors related to the copying and pasting of code.

Download the example code files

You can download the example code files for this book from GitHub at `https://github.com/PacktPublishing/Modern-Frontend-Development-with-Node.js`. If there's an update to the code, it will be updated in the GitHub repository.

We also have other code bundles from our rich catalog of books and videos available at `https://github.com/PacktPublishing/`. Check them out!

Code in Action

The Code in Action videos for this book can be viewed at `http://bit.ly/3EgcKwM`.

Download the color images

We also provide a PDF file that has color images of the screenshots and diagrams used in this book. You can download it here: `https://packt.link/zqKz4`.

Conventions used

There are a number of text conventions used throughout this book.

`Code in text`: Indicates code words in text, database table names, folder names, filenames, file extensions, pathnames, dummy URLs, user input, and Twitter handles. Here is an example: "Mount the downloaded `WebStorm-10*.dmg` disk image file as another disk in your system."

A block of code is set as follows:

```
html, body, #map {
 height: 100%;
 margin: 0;
 padding: 0
}
```

When we wish to draw your attention to a particular part of a code block, the relevant lines or items are set in bold:

```
[default]
exten => s,1,Dial(Zap/1|30)
exten => s,2,Voicemail(u100)
exten => s,102,Voicemail(b100)
exten => i,1,Voicemail(s0)
```

Any command-line input or output is written as follows:

```
$ npm install
```

Bold: Indicates a new term, an important word, or words that you see onscreen. For instance, words in menus or dialog boxes appear in **bold**. Here is an example: "Select **System info** from the **Administration** panel."

> **Tips or important notes**
> Appear like this.

Get in touch

Feedback from our readers is always welcome.

General feedback: If you have questions about any aspect of this book, email us at customercare@packtpub.com and mention the book title in the subject of your message.

Errata: Although we have taken every care to ensure the accuracy of our content, mistakes do happen. If you have found a mistake in this book, we would be grateful if you would report this to us. Please visit www.packtpub.com/support/errata and fill in the form.

Piracy: If you come across any illegal copies of our works in any form on the internet, we would be grateful if you would provide us with the location address or website name. Please contact us at copyright@packt.com with a link to the material.

If you are interested in becoming an author: If there is a topic that you have expertise in and you are interested in either writing or contributing to a book, please visit authors.packtpub.com.

Share your thoughts

Once you've read *Modern Frontend Development with Node.js*, we'd love to hear your thoughts! Scan the QR code below to go straight to the Amazon review page for this book and share your feedback.

https://packt.link/r/1-804-61829-2

Your review is important to us and the tech community and will help us make sure we're delivering excellent quality content.

Download a free PDF copy of this book

Thanks for purchasing this book!

Do you like to read on the go but are unable to carry your print books everywhere?

Is your eBook purchase not compatible with the device of your choice?

Don't worry, now with every Packt book you get a DRM-free PDF version of that book at no cost.

Read anywhere, any place, on any device. Search, copy, and paste code from your favorite technical books directly into your application.

The perks don't stop there, you can get exclusive access to discounts, newsletters, and great free content in your inbox daily!

Follow these simple steps to get the benefits:

1. Scan the QR code or visit the link below:

https://packt.link/free-ebook/9781804618295

2. Submit your proof of purchase
3. That's it! We'll send your free PDF and other benefits to your email directly

Part 1:
Node.js Fundamentals

In this part, you'll dive into Node.js by learning how it works and how you can use it. You'll also get in touch with the Node.js ecosystem. In particular, you'll see how Node.js projects are structured. An important topic of this part is how to deal with dependencies in the form of packages.

This part of the book comprises the following chapters:

- *Chapter 1, Learning about the Internals of Node.js*
- *Chapter 2, Dividing Code into Modules and Packages*
- *Chapter 3, Choosing a Package Manager*

1
Learning about the Internals of Node.js

For years, being a frontend developer meant writing a bit of HTML and putting some styling with CSS on it. However, since the last decade, this job description barely holds true. In contrast, the majority of frontend work is now done using **JavaScript**.

Initially used to make cosmetic enhancements to websites (such as the toggling of elements) possible, frontend development is now the glue of the web. Websites are no longer just written in HTML and CSS. Instead, in many cases, web pages are programmed with JavaScript using modern techniques such as dependency management and bundling of resources. The **Node.js** framework provides an ideal foundation for this movement. It enables developers to use JavaScript not only inside websites running in a browser but also within the tooling to write web pages – outside of a browser.

When Node.js was released in May 2009, it did not seem like a big deal. JavaScript was working on the server too. However, the cross-platform nature of Node.js and the size of the JavaScript community provided the basis for one of the greatest disruptions in the history of computing. People started adopting the framework so quickly that many existing frameworks either disappeared or had to be reworked to stay attractive to developers. Soon, JavaScript was used in the browser and on the server and was also part of every frontend developer's toolbox.

With the rise of new development frameworks such as **Angular** or **React**, the need for attractive frontend tooling became apparent. The new frameworks always relied on some build steps – otherwise, websites and applications using these frameworks would have been far too inconvenient to write for developers. Since the vast Node.js ecosystem seemed to have figured out a suitable approach for reusability, these new frameworks adopted it and made it an integral part of their development story. This way, using Node.js became the de facto standard for frontend projects of any kind.

Today, it is pretty much impossible to start a frontend development project without having Node.js installed. In this book, we'll take the journey of learning about Node.js from the inside out together. We will not be focusing on writing server applications or walking over the integrated functionality of Node.js. Instead, we'll look at how we – as frontend developers – can leverage the best that Node.js brings to the table.

In this first chapter, we discuss the internals of Node.js. This will help you understand how Node.js works and how you can actually use it. After this chapter, you will be able to run and debug simple scripts using the Node.js command-line application.

We will cover the following key topics in this chapter:

- Looking at the Node.js architecture in detail
- Understanding the event loop
- Using Node.js from the command line
- CommonJS

Technical requirements

To follow the code samples in this book, you need knowledge of JavaScript and how to use the command line. You should have Node.js installed using the instructions at `https://nodejs.org`.

The complete source code for this chapter is available at `https://github.com/PacktPublishing/Modern-Frontend-Development-with-Node.js/tree/main/Chapter01`.

The Code in Action (CiA) videos for this chapter can be accessed at `http://bit.ly/3fPPdtb`.

Looking at the Node.js architecture in detail

The principal foundations of Node.js have been inspired by a few things:

- The single worker thread featured in browsers was already quite successful in the server space. Here, the popular **nginx** web server showed that the event loop pattern (explained later in this chapter) was actually a blessing for performance – eliminating the need to use a dedicated thread pool for handling requests.
- The idea of packaging everything in a file-centric structure called **modules**. This allowed Node.js to avoid many of the pitfalls of other languages and frameworks – including JavaScript in the browser.
- The idea of avoiding creating a huge framework and leaving everything extensible and easy to get via package managers.

Threads

Modern computers offer a lot of computing power. However, for an application to really use the available computing power, we need to have multiple things working in parallel. Modern operating systems know about different independently running tasks via so-called threads. A **thread** is a group of operations running sequentially, which means in a given order. The operating system then schedules when threads run and where (i.e., on which CPU core) they are placed.

These principles together form a platform that seems easy to create, but hard to replicate. After all, there are plenty of JavaScript engines and useful libraries available. For Ryan Dahl, the original creator and maintainer of Node.js, the basis of the framework had to be rock solid.

Ryan Dahl selected an existing JavaScript engine (**V8**) to take over the responsibility of parsing and running the code written in JavaScript. The V8 engine was chosen for two good reasons. On the one hand, the engine was available as an open source project under a permissive license – usable by projects such as Node.js. On the other hand, V8 was also the engine used by Google for its web browser **Chrome**. It is very fast, very reliable, and under active development.

One of the drawbacks of using V8 is that it was written in C++ using custom-built tooling called **GYP**. While GYP was replaced in V8 years later, the transition was not so easy for Node.js. As of today, Node.js is still relying on GYP as a build system. The fact that V8 is written in C++ seems like a side note at first, but might be pretty important if you ever intend to write so-called **native modules**.

Native modules allow you to go beyond JavaScript and Node.js – making full use of the available hardware and system capabilities. One drawback of native modules is that they must be built on each platform. This is against the **cross-platform** nature of Node.js.

Let's take a step back to arrange the parts mentioned so far in an architecture diagram. *Figure 1.1* shows how Node.js is composed internally:

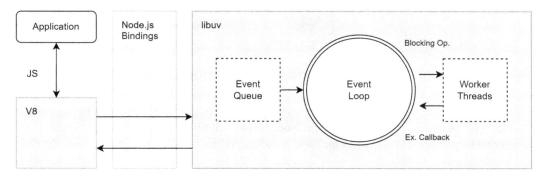

Figure 1.1 – Internal composition of Node.js

The most important component in Node.js's architecture – besides the JavaScript engine – is the **libuv** library. libuv is a multi-platform, low-level library that provides support for asynchronous **input/output (I/O)** based on an **event loop**. I/O happens in multiple forms, such as writing files or handling **HTTP requests**. In general, I/O refers to anything that is handled in a dedicated area of the operating system.

Any application running Node.js is written in JavaScript or some flavor of it. When Node.js starts running the application, the JavaScript is parsed and evaluated by V8. All the standard objects, such as console, expose some bindings that are part of the Node.js API. These low-level functions (such as console.log or fetch) make use of libuv. Therefore, some simple script that only works against language features such as primitive calculations (*2 + 3*) does not require anything from the Node API and will remain independent of libuv. In contrast, once a low-level function (for example, a function to access the network) is used, libuv can be the workforce behind it.

In *Figure 1.2*, a block diagram illustrating the various API layers is shown. The beauty of this diagram is that it reveals what Node.js actually is: a JavaScript runtime allowing access to low-level functionality from state-of-the-art **C/C++** libraries. The Node.js API consists of the included Node.js bindings and some C/C++ addons:

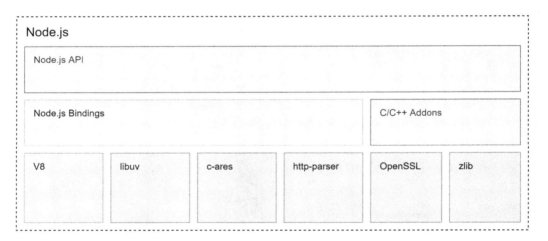

Figure 1.2 – Composition of Node.js in terms of building blocks

One thing that would need explanation in the preceding diagram is how the event loop is implemented in relation to all the blocks. When talking about Node.js's internal architecture, a broader discussion of what an event loop is and why it matters for Node.js is definitely required. So let's get into these details.

Understanding the event loop

An event loop is a runtime model that enables users to run all operations from a single thread – irrespective of whether the operations access long-running external resources or not. For this to work, the event loop needs to make requests to an event provider, which calls the specified event handlers. In Node.js, the libuv library is used for event loop implementation.

The reason for giving libuv the most space in *Figure 1.1* is to highlight the importance of this library. Internally, libuv is used for everything regarding I/O, which arguably is the most crucial piece of any framework. I/O lets a framework communicate with other resources, such as files, servers, or databases. By default, dealing with I/O is done in a blocking manner. This means that the sequence of operations in our application is essentially stopped, waiting for the I/O operation to finish.

Two strategies for mitigating the performance implications of blocking I/O exist.

The first strategy is to create new threads for actually performing these blocking I/O operations. Since a thread contains an independent group of operations, it can run concurrently, eventually not stopping the operations running in the original thread of the application.

The second strategy is to not use **blocking I/O** at all. Instead, use an alternative variant, which is usually called non-blocking I/O or asynchronous I/O. **Non-blocking I/O** works with callbacks, that is, functions that are called under certain conditions – for instance when the I/O operation is finished. Node.js uses libuv to make extensive use of this second strategy. This allows Node.js to run all code in a single thread, while I/O operations run concurrently.

In *Figure 1.3*, the building blocks of libuv are displayed. The key part is that libuv already comes with a lot of functionality to handle network I/O. Furthermore, file and DNS operations are also covered well:

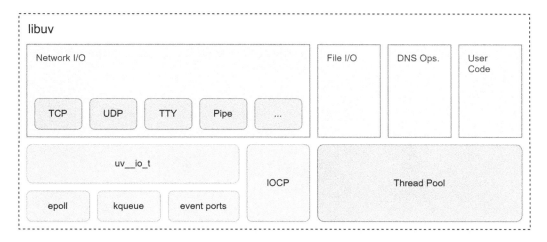

Figure 1.3 – Building blocks of libuv

In addition to the different I/O operations, the library comes with a set of different options for handling asynchronous user code.

The event loop itself follows the **reactor design pattern**. Wikipedia describes the pattern as follows:

> *The reactor design pattern is an event handling pattern for handling service requests delivered concurrently to a service handler by one or more inputs. The service handler then demultiplexes the incoming requests and dispatches them synchronously to the associated request handlers. (https://en.wikipedia.org/wiki/ Reactor_pattern)*

Importantly, this definition mentions synchronous dispatch. This means that code that is run through the event loop is guaranteed to not run into any conflicts. The event loop makes sure that code is always run sequentially. Even though the I/O operations may concurrently run, our callbacks will never be invoked in parallel. From our perspective, even though Node.js will internally (through libuv) use multiple threads, the whole application is single-threaded.

The following is a simple script that shows you the basic behavior of the event loop at play – we'll discuss how to run this in the *Using Node.js from the command line* section:

events.js

```
console.log('A [start]');
setTimeout(() => console.log('B [timeout]'), 0);
Promise.resolve().then(() => console.log('C [promise]'));
console.log('D [end]');
```

We will run this script in the next section when we learn about the command line usage of Node. js. In the meantime, put some thought into the preceding code and write down the order in which you'll see the console output. Do you think it will print in an "A B C D" order, or something else?

The algorithm of the implementation of the event loop in libuv is displayed in *Figure 1.4*:

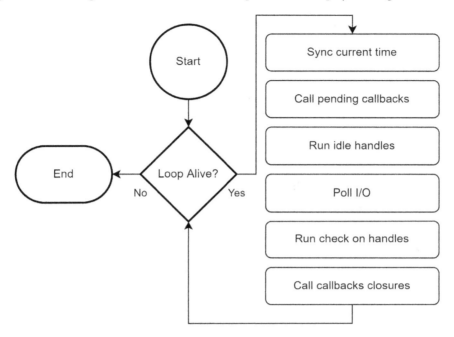

Figure 1.4 – The implementation of the event loop in libuv

While the code snippet only deals with JavaScript-related constructs (such as `console`, `Promise`, and `setTimeout`), in general, the callbacks are associated with resources that go beyond Node. js, such as file system changes or network requests. Some of these resources may have an operating system equivalent; others only exist in form of blocking I/O.

Consequently, the event loop implementation always considers its thread pool and polls for progressed I/O operations. Timers (such as `setTimeout` in the example script) are only run in the beginning. To know whether a timer needs to be run, its due time is compared with the current time. The current time is synced with the system time initially. If there is nothing to be done anymore (that is, no active timer, no resource waiting to finish, etc.), then the loop exits.

Let's see how we can run Node.js to solidify our knowledge about the event loop.

Using Node.js from the command line

Using JavaScript for a web application just requires you to open the website in a browser. The browser will evaluate the included JavaScript and run it. When you want to use JavaScript as a scripting language, you need to find a new way of running JavaScript. Node.js offers this new way – running JavaScript in a terminal, inside our computer, or from a server.

When Node.js is installed, it comes with a set of command-line tools that will be available in the terminal of your choice. For this book, you'll need to know about three different executables that we'll use throughout the chapters:

- **node**: The main application to run a Node.js script
- **npm**: The default package manager – more on that later
- **npx**: A very convenient utility to run npm binaries

For now, we only need to know about node. If we want to run the events.js script from the previous section, we need to execute the following command in the directory in which the script (events.js) has been placed. You can place it there by just inserting the content from the previous events.js listing:

```
$ node events.js
A [start]
D [end]
C [promise]
B [timeout]
```

The command is shown after the conventional $ sign indicating the command prompt. The output of running the script is shown below the node events.js command.

As you can see, the order is "A D C B" – that is, Node.js first handled all the sequential operations before the callbacks of the promise were handled. Finally, the timeout callback was handled.

The reason for handling the promise callback before the timeout callback lies in the event loop. In JavaScript, promises spawn so-called micro tasks, which are placed in the pending callback section of the libuv event loop from *Figure 1.4*. The timeout callback, however, is treated like a full task. The difference between them lies within the event loop. Micro tasks are placed in an optimized queue that is actually peeked multiple times per event loop iteration.

According to libuv, the timeout callback can only be run when its timer is due. Since we only placed it in the event loop during the idle handles (i.e., main section) of the event loop, we need to wait until the next iteration of the event loop.

The `node` command-line application can also receive additional parameters. The official documentation goes into all details (`https://nodejs.org/api/cli.html`). A helpful one is `-e` (short version of `--eval`) to just evaluate a script directly from the command-line input without requiring a file to run:

```
$ node -e "console.log(new Date())"
2022-04-29T09:20:44.401
```

Another very helpful command line flag is `--inspect`. This opens the standard port for graphical inspection, for example, via the Chrome web browser.

Let's run an application with a bit of continuous logic to justify an inspection session. In the terminal on your machine, run the following:

```
$ node -e "setInterval(() => console.log(Math.random()), 60 *
1000)" --inspect
Debugger listening on ws://127.0.0.1:9229/64c26b8a-0ba9-484f-
902d-759135ad76a2
For help, see: https://nodejs.org/en/docs/inspector
```

Now we can run a graphical application. Let's use the Chrome web browser. Open it and go to `chrome://inspect`. This is a special Chrome-internal URL that allows us to see the available targets.

The following figure (*Figure 1.5*) shows how inspecting the Node.js application in the Chrome web browser may look:

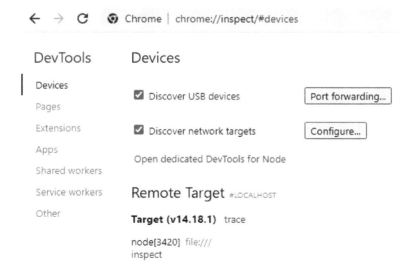

Figure 1.5 – Inspecting the Node.js application in the Chrome web browser

In this case, Chrome detected our application with the process ID 3420 running. On your machine, the process ID will most likely be different. No filename was given, as we started with the -e command-line option.

When you click on **inspect**, you'll open the standard Chrome **DevTools**. Instead of debugging a website, you can now debug the Node.js application. For instance, you'll already get the same console output that you see in the command line.

When you follow the link to the evaluated script from the DevTools console, you'll get the ability to place **breakpoints** or pause the execution. Pausing the execution may not work immediately, as an active JavaScript operation is required for that.

In *Figure 1.6*, you see how debugging a Node.js script in the Chrome DevTools can look:

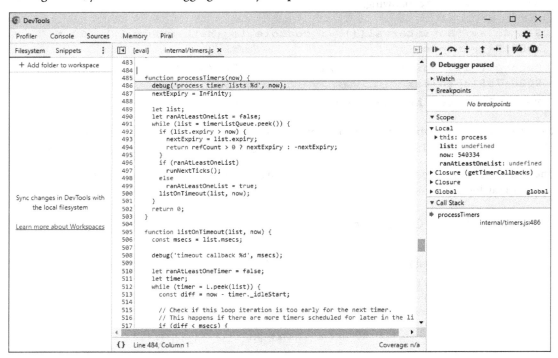

Figure 1.6 – Debugging a Node.js script in the Chrome DevTools

In the preceding example, JavaScript is only run every minute. When the pause occurs, you should end up in the internal/timers.js part of Node.js itself. This is a different JavaScript file, but it's part of the whole Node.js framework. The file can be integrated because it follows certain conventions and rules that are known as CommonJS.

CommonJS

One thing that Node.js got right from the beginning was to introduce an explicit way of obtaining and using functionality. JavaScript in the browser suffered from the *global scope* problem, which caused many headaches for developers.

> **Global scope**
>
> In JavaScript, the global scope refers to functionality that is accessible from every script running in the same application. On a website, the global scope is usually the same as the window variable. Attaching variables to the global scope may be convenient and sometimes even necessary, but it may also lead to conflicts. For instance, two independent functions could both try to write and read from the same variable. The resulting behavior can then be hard to debug and very tricky to resolve. The standard recommendation is to avoid using the global scope as much as possible.

The idea that other functionalities are explicitly imported was certainly not new when Node.js was introduced. While an import mechanism existed in other programming languages or frameworks for quite some time, similar options have also been available for JavaScript in the browser – via third-party libraries such as **RequireJS**.

Node.js introduced its **module system** with the name **CommonJS**. The basis for Node.js's implementation was actually a project developed at Mozilla. In that project, Mozilla worked on a range of proposals that started with non-browser use but later on expanded to a generic set of JavaScript specifications for a module system.

> **CommonJS implementations**
>
> Besides the implementation in Node.js, many other runtimes or frameworks use CommonJS. As an example, the JavaScript that can be used in the **MongoDB** database leverages a module system using the CommonJS specifications. The implementation in Node.js is actually only partially fulfilling the full specification.

A module system is crucial for allowing the inclusion of more functionality in a very transparent and explicit manner. In addition to a set of more advanced functionality, a module system gives us the following:

- A way of including more functionality (in CommonJS, via the global require function)
- A way of exposing functionality, which can then be included somewhere else (in CommonJS, via the module-specific module or exports variables)

At its core, the way CommonJS works is quite simple. Imagine you have a file called a.js, which contains the following code:

```
const b = require('./b.js');
console.log('The value of b is:', b.myValue);
```

Now the job of Node.js would be to actually make this work, that is, give the b variable a value that represents the so-called exports of the module. Right now, the script would error out saying that a b.js file is missing.

The b.js file, which should be adjacent to a.js, reads as follows:

```
exports.myValue = 42;
```

When Node.js evaluates the file, it will remember the defined exports. In this case, Node.js will know that b.js is actually just an object with a myValue key with a value of 42.

From the perspective of a.js, the code can therefore be read like this:

```
const b = {
  myValue: 42,
};
console.log('The value of b is:', b.myValue);
```

The advantage of using the module system is that there is no need to write the outputs of the module again. The call to require does that for us.

Side effects

Replacing the call to require with the module's outputs is only meant for illustrative purposes. In general, this cannot be done as the module evaluation can have some so-called side effects. A **side effect** happens when implicit or explicit global variables are manipulated. For instance, already writing something to the console or outputting a file in the module evaluation is a side effect. If we'd only replace the require call with the imported module's exports, we would not run the side effects, which would miss a crucial aspect of the module.

In the given example, we used the name of the file directly, but importing a module can be more subtle than that. Let's see a refined version of the code:

---

a.js

```
const b = require('./b');
console.log('The value of b is:', b.myValue);
```

The call to ./b.js has been replaced by ./b. This will still work, as Node.js will try various combinations for the given import. Not only will it append certain known extensions (such as .js) but it will also look at whether b is actually a directory with an index.js file.

Therefore, with the preceding code, we could actually move b.js from a file adjacent to a.js to an index.js file in the adjacent directory, b.

The greatest advantage, however, is that this syntax also allows us to import functionality from third-party packages. As we will explore later in *Chapter 2, Dividing Code into Modules and Packages*, our code has to be divided into different modules and packages. A **package** contains a set of reusable modules.

Node.js already comes with a set of packages that don't even need to be installed. Let's see a simple example:

host.js

```
const os = require('os');
console.log('The current hostname is:', os.hostname());
```

The preceding example uses the integrated os package to obtain the current computer's network name.

We can run this script with node in the command line:

```
$ node host.js
The current hostname is: DESKTOP-3JMIDHE
```

This script works on every computer that has Node.js installed.

Summary

In this chapter, we discovered Node.js for the first time. You should now have a good idea of the core principles (such as event loop, threads, modules, and packages) upon which Node.js was built. You have read a bit about Node.js's history and why V8 was chosen as the JavaScript engine.

One of the key things to take away from this chapter is how the event loop works. Note that part of this knowledge is not exclusive to Node.js. The distinction between micro tasks and tasks is an integral part of how JavaScript engines, even the JavaScript engine of your browser, work.

Lastly, you are now equipped to use the node command-line application, for example, to run or debug simple scripts, which can export and import functionality using the CommonJS module system. You learned how to use the Chrome web browser to inspect Node.js scripts as you can with websites.

In the next chapter, we will increase our knowledge about CommonJS by learning how we can efficiently divide code into modules and packages.

2
Dividing Code into Modules and Packages

One of the most important aspects to consider when writing professional software is reusability. Reusability means that parts of our code base can be purposed to work in several places or under different circumstances. This implies that we can actually use existing functionality quite easily.

As we learned, a key part of the Node.js success story is down to the fact that it comes with a module system. So far, we've only touched upon the basic concept of CommonJS, which is the default way of importing and exporting functionality from modules.

In this chapter, we'll take the chance to become familiar with more module formats, including their history, use cases, and development models. We'll learn how to divide our code into modules and packages efficiently. In addition to learning about CommonJS, we will see what a package is and how we can define our own packages. All in all, this will help us to achieve great reusability – not only for our tooling in Node.js but also for our applications running in the browser.

We will cover the following key topics in this chapter:

- Using the ESM standard
- Learning the AMD specification
- Being universal with UMD
- Understanding SystemJS and import maps
- Knowing the `package.json` fundamentals

Technical requirements

The complete source code for this chapter can be found at `https://github.com/PacktPublishing/Modern-Frontend-Development-with-Node.js/tree/main/Chapter02`.

The CiA videos for this chapter can be accessed at `http://bit.ly/3FZ6ivk`.

Using the ESM standard

CommonJS has been a good solution for Node.js, but not a desirable solution for JavaScript as a language. For instance, in the browser, CommonJS does not work. Doing synchronous imports on URLs is just not possible. The module resolution of CommonJS was also way too flexible in terms of adding extensions and trying directories.

To standardize modules in JavaScript, the **ECMAScript Module** (**ESM**) standard was established. It is capable of defining modules that run in the browser, as well as Node.js. Furthermore, instead of using an arbitrary function such as `require`, the whole module system relies on language constructs using reserved words. This way, the module system can be brought over to the browser, too.

The ECMAScript standard specified two keywords for this:

- `import`: Used to import functionality from other modules
- `export`: Used to declare the functionality that can be imported into other modules

The `import` keyword must appear at the beginning of a file – before any other code. The reason for this choice lies in the demand for ESM files to be used not only within Node.js, but also in the browser. By placing the `import` statements on top, each ESM file can safely wait until all the imports have been resolved.

Rewriting the example from the previous chapter, we get the following for `a.js`:

```
import * as b from './b.js'; // get all things from b.js
// use imports
console.log('The value of b is:', b.myValue);
```

The rewrite of the `b.js` file to be valid per the ESM standard is as follows:

```
export const myValue = 42;
```

There are multiple possibilities with the `import` keyword. We can use the following:

- Wildcard (using `*`) imports with a name selected by the developer
- Named imports such as `myValue`

- Default imports with a name selected by the developer
- An empty import that does not get anything, but makes sure to run the module

Using a named import, we can get a cleaner version of a.js:

```
// get only selected things
import { myValue } from './b.js';
console.log('The value of b is:', myValue); // use imports
```

The preceding code is very similar to the destructuring assignment, which decomposes an object into its fields using the assignment operator (=). There are crucial differences, however. One of these differences is how to make aliases.

For instance, when using a destructuring assignment, we can use the colon (:) to rename the variables, which would have the name of the respective fields by default. If we wanted to give the variable a different name (e.g., otherValue) from its original field (e.g., myValue), we'd have to write the following:

```
// gets all the things, but only uses myValue
const { myValue: otherValue } = require('./b.js');
```

With an import statement, you need to use the as keyword to achieve this:

```
// gets only myValue - but renames it
import { myValue as otherValue } from './b.js';
```

A topic that becomes relevant quite quickly is the notion of a default export. Especially when handling exports from an unknown module, there is a great need to define the export name. In CommonJS, developers therefore picked the whole module; however, this is no longer possible with ESM. Every export needs to be named.

Luckily, the standardization committee thought about the topic of default exports. An export is considered to be a default export if it uses the default keyword. For instance, changing the export in b.js to use default values could look as follows:

```
export default 42;
```

Importing the default export is quite convenient, too. Here, we are free to select a name to refer to the default export within our module. Instead of being able to rename the import, we are forced to give it a name:

```
import otherValue from './b.js'; // gets only default
console.log('The value of b is:', otherValue);
```

The whole idea is to use default exports as much as possible. In the end, modules that are effectively written to revolve around exporting a single functionality are often considered the goal.

We've already learned that CommonJS does not work in the browser. In contrast, the modern ESM specification is supposed to work, as imports are declared in the beginning. This modification allows the browser to safely suspend module evaluation until the imports are fully processed. This kind of suspension to wait for the dependencies to finish loading was actually taken from another attempt at a module system called **Asynchronous Module Definition (AMD)**.

Learning the AMD specification

Before ESM was established, people tried to make modules work in the browser, too. One of the earliest attempts was a small library called **RequireJS**. RequireJS is a module loader that works in the browser as well as in Node.js. Initially, the essential idea was that a script reference to RequireJS would be embedded in the <head> of a document. The script would then load and run a defined root module, which would process even more modules.

An example website using RequireJS is as follows:

```
<!DOCTYPE html>
<html>
  <head>
    <title>My Sample Project</title>
    <!--
      data-main attribute tells RequireJS to load
      ./main.js after ./require.js has been loaded
    -->
    <script data-main="./main" src="./require.js"></script>
  </head>
  <body></body>
</html>
```

RequireJS was born at a time when promises had not yet been established in the JavaScript world. Therefore, the module loader was based on the next best thing: callbacks. Consequently, a module is loaded by calling a `requirejs` function defined by RequireJS. The whole process can then start loading modules asynchronously as shown in *Figure 2.1*:

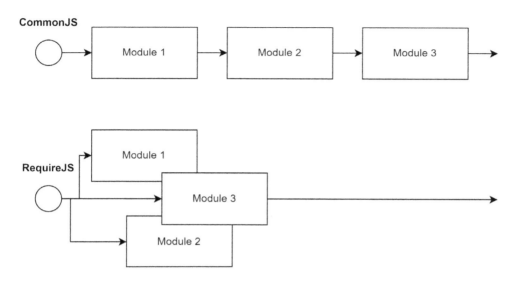

Figure 2.1 – Loading modules sequentially vs. asynchronously

To achieve asynchronous module loading, the `requirejs` function takes two arguments. The first argument is an array with all the dependencies. The second argument is a callback that receives the exports of the dependencies and returns the exports of the current module.

The whole idea behind RequireJS is very similar to that of ESM today, which shifts the two parts (loading the dependencies and the code that uses the dependencies) into the same module – but still distinguishes between the `import` statements and all the other statements. Here, ESM leverages the fact that it's actually a language construct.

In short, a module that uses RequireJS looks as follows:

```
requirejs(['./helper/util'], (util) => {
  // This is called when ./helper/util.js. has been processed
});
```

The shape of these modules was not arbitrarily decided. Instead, the RequireJS library is merely one implementation of a specification for an asynchronous module system. This specification is called AMD.

Using AMD, the previous RequireJS-specific example could be rewritten as follows:

```
define(['./helper/util'], (util) => {
  // This is called when ./helper/util.js. has been processed
});
```

In addition to the two-argument variant of the `define` function, there is also a three-argument version, where the first argument helps to name the defined module.

An example of the three-argument call to `define` is shown here:

```
define('myModule', ['dep1', 'dep2'], (dep1, dep2) => {
  // Define the module exports by returning a value.
  return {};
});
```

Now, the only thing left before we can use AMD universally is to learn how we can integrate it into Node.js. First, we need to grab `r.js` from the official download page: `https://requirejs.org/docs/download.html#rjs`. Download it via the **Download** button as shown in *Figure 2.2*:

r.js: Optimizer and Node/Rhino/Nashorn/xpcshell adapter

The r.js file allows you to run the optimizer as well as run modules in Node, Rhino, Nashorn or xpcshell.

Figure 2.2 – The Download button for r.js on the RequireJS website

Store the downloaded file in the directory where you have placed the scripts to run via `node`. Create a new `a.js` script in the same directory:

a.js

```
const define = require('./r.js'); // gets the loader

define.config({
  // Will also correctly resolve other Node.js dependencies
  nodeRequire: require
});

define(['./b'], (b) => {
  console.log('The value of b is:', b.myValue);
});
```

The code here does not look so different from the CommonJS example. After all, only the initialization of the RequireJS loader has been added. The actual content of the module is now part of the callback.

Let's see what the transformed b.js looks like:

b.js

```
const define = require('./r.js'); // gets the loader

define.config({
  // Will also correctly resolve other Node.js dependencies
  nodeRequire: require
});

define([], () => {
  return {
    myValue: 42,
  };
});
```

In the preceding code for b.js, again, we have added the same envelope, just as in a.js. Remember that each module needs to be treated as standalone code. While how explicit this is may seem rather redundant, the real advantage becomes obvious once it's used with an unknown number of other modules. In this case, we never know what has been loaded or used already. Being independent means being predictable in these scenarios.

The problem with the preceding approach is that while this works in Node.js, it certainly does not work in the browser. Even though we've chosen AMD for this specific reason, we failed to make it work in the browser. The problem lies in the initial call to require, which uses CommonJS to obtain the AMD loader.

To mitigate the problem and use AMD in different JavaScript environments, the **Universal Module Definition** (**UMD**) specification was created.

Being universal with UMD

When the UMD specification was brought up, there was a lot of hype in the community. After all, the label *universal* already claims that UMD is the final module system – the one to rule them all. It tries to do this by supporting essentially three different kinds of JavaScript module formats:

- The classic way of doing things without a module system – that is, just by running JavaScript using <script> tags in the browser

- The CommonJS format that is used by Node.js

- The previously discussed asynchronously loaded modules from the AMD specification

When you write a JavaScript file with the UMD specification in mind, you essentially make sure that every popular JavaScript runtime can read it. For instance, UMD works perfectly in Node.js and the browser.

To achieve this universality, UMD makes an educated guess regarding what module system can be used and selects it. For example, if a `define` function is detected, then AMD might be used. Alternatively, detecting something such as `exports` or `module` hints towards CommonJS. If nothing is found, then the assumption is that the module runs in the browser without AMD present. In this case, the exports of the module would be stored globally.

The main target group for UMD is library authors. When you build a library, you want it to be useful. Consequently, you'll also need to make sure that the library can be used. By providing your library in the UMD format, you ensure that it can be used on pretty much all platforms – in Node.js and the browser.

So, how would our code from the previous example look if we used UMD as the format of choice? Let's have a look:

a.js

```
((root, factory) => { // context and export callback
  if (typeof define === 'function' && define.amd) {
    // there is a define function that follows AMD - use it
    define(['b'], factory);
  } else if (typeof exports === 'object' && typeof module
  !== 'undefined') {
    // there is module and exports: CommonJS
    factory(require('b'));
  } else {
    // we just take the global context
    factory(root.b);
  }
}) (typeof self !== 'undefined' ? self : this, (b) => {
  // this is the body of the module, follows AMD
  console.log('The value of b is:', b.myValue);
});
```

As before, the preceding code consists of two sections. The first section establishes the module system and sets up the callback. The second section puts the actual content of our module into the callback.

The only thing left is to see how we can mark our exports with UMD. For this part, we will look at b.js in the UMD format:

b.js

```
((root, factory) => {
  if (typeof define === 'function' && define.amd) {
    // in AMD we depend on the special "exports" dependency
    define(['exports'], factory);
  } else if (typeof exports === 'object' && typeof module
  !== 'undefined') {
    // in CommonJS we'll forward the exports
    factory(exports);
  } else {
    // for scripts we define a new global and forward it
    factory(root.b = {});
  }
})(typeof self !== 'undefined' ? self : this, (exports) =>
{
  // use the CommonJS format in here
  exports.myValue = 42;
});
```

With all the boilerplate code in place, the script is universal. The defined callback (named factory in the two examples in this section) is either called indirectly from the AMD runtime or directly in the other two cases.

Usually, we will not write the whole boilerplate code shown here ourselves. The boilerplate will be generated by tooling, which we will look into in *Chapter 6, Building Web Apps with Bundlers*. Instead, the ideal option for writing modules in many cases is ESM. Since it's syntax-based, we follow the language's standard. The other formats can then be used by our tooling as output formats.

One more module format to have a closer look at is SystemJS. One of the reasons why SystemJS is interesting is that it brings support for import maps, which can simplify dealing with module systems.

Understanding SystemJS and import maps

Earlier in this chapter, we learned that ESM is arguably the best module system for JavaScript. After all, it is integrated into the JavaScript language. One of the reasons why other formats are still relevant today is backward compatibility.

Backward compatibility allows formats such as AMD or UMD to be used in older JavaScript runtimes, such as older versions of browsers such as Internet Explorer, but even if we don't need backward compatibility, the alternative formats still have one or more advantages over ESM.

One of the core problems with ESM is that it does not define how modules are resolved. In fact, the only specified way to resolve a module is explicitly via the filesystem. When we used ESM, we explicitly stated our module imports, such as in ./b.js. As mentioned, we are not allowed to implicitly use something such as ./b or even just b.

When doing frontend development, the notion of dependencies has become quite elementary. From boilerplate libraries to UI frameworks, frontend developers make use of a wide array of given code. That code is usually packaged into libraries and then installed locally for development purposes, but how should these dependencies be used?

Turns out that Node.js solved this problem already in the early stages of its development. We have seen that using CommonJS we are able to write code such as the following:

host-cjs.js

```
const os = require('os');
console.log('The current hostname is:', os.hostname());
```

The reference to os is resolved by Node.js through CommonJS. In this special case, the reference leads to one framework library of Node.js. However, it could also lead to a third-party dependency that has been installed by us. In *Chapter 3*, *Choosing a Package Manager*, we will see how this works.

Let's translate the preceding code into ESM:

host-esm.js

```
import { hostname } from 'node:os';
console.log('The current hostname is:', hostname());
```

The conversion of the little snippet is not very complicated, with the exception of the module name. Previously, we used os as an identifier. Node.js has chosen to also allow this for backward compatibility – at least for now. The preferred way, however, is to use a custom protocol. In the case of Node.js framework libraries, the node: protocol has been chosen.

Leveraging custom protocols to resolve dependencies is possible in the browser. However, it is also cumbersome. After all, the whole resolution would now need to be done by us. This also represents a classic chicken-egg problem. To define custom protocols, we need to have some JavaScript running; however, if this piece of JavaScript relies on third-party dependencies that are actually resolved via the custom protocol, then we cannot successfully implement the resolution of dependencies.

One way that we can still use convenient references such as `os` is to define a so-called import map. An import map helps the browser map module names to actual URLs. It uses **JSON** with an object stored in the `imports` field.

The following is an import map to find an implementation of the `os` module:

```json
{
  "imports": {
    "os": "https://example.com/js/os.min.js"
  }
}
```

The URLs don't have to be fully qualified. In the case of relative URLs, the module's URL is computed from the base URL of the import map.

The integration of import maps into a website is relatively simple. All we need to do is to specify a `<script>` tag with the type being `importmap`:

```html
<script type="importmap">
{
  "imports": {
    "os": "https://example.com/js/os.min.js"
  }
}
</script>
```

In addition, import maps may be loaded from external files, too. In any case, the specified mapping of module names to URLs only works for `import` statements. It will not work in other places where a URL is expected. For instance, the following example does not work:

fail.html

```html
<script type="importmap">
{
  "imports": {
    "/app.mjs": "/app.8e0d62a03.mjs"
```

```
    }
}
</script>
<script type="module" src="/app.mjs"></script>
```

In the preceding code, we have tried to load /app.mjs directly, which will fail. We need to use an import statement:

success.html

```
<script type="importmap">
{
  "imports": {
    "/app.mjs": "/app.8e0d62a03.mjs"
  }
}
</script>
<script type="module">import "/app.mjs";</script>
```

There is a lot more that can be written about import maps; however, for now, the most important detail is that they only work partially – that is, without external files, in recent versions of *Google Chrome* (*89* and higher) and *Microsoft Edge* (*89* and higher). In most other browsers, the import map support is either not there or must explicitly be enabled.

The alternative is to use SystemJS. SystemJS is a module loader similar to RequireJS. The main difference is that SystemJS provides support for multiple module systems and module system capabilities, such as using import maps.

While SystemJS also supports various formats such as ESM, it also comes with its own format. Without going into too much detail, the shape of a native SystemJS module looks as follows:

```
System.register(['dependency'], (_export, _context) => {
  let dependency;
  return {
    setters: [(_dep) => {
      dependency = _dep;
    }],
    execute: () => {
      _export({
        myValue: 42,
```

```
        });
      },
    };
});
```

The preceding code is structurally quite similar to the AMD boilerplate, with the only difference being how the callback is structured. While AMD runs the module's body in the callback, SystemJS specifies some more sections in the callback. These sections are then run on demand. The real body of a module is defined in the returned `execute` section.

As before, the short snippet already illustrates quite nicely that SystemJS modules are rarely written by hand. Instead, they are generated by tooling. We'll therefore come back to SystemJS once we have more powerful tooling on hand to automate the task of creating valid SystemJS modules.

Now that we have heard enough about libraries and packages, we also need to know how we can define our own package. To indicate a package, the `package.json` file has to be used.

Knowing package.json fundamentals

The aggregation of multiple modules forms a package. A package is defined by a `package.json` file in a directory. This marks the directory as the root of a package. A minimal valid `package.json` to indicate a package is as follows:

package.json

```
{
  "name": "my-package",
  "version": "1.0.0"
}
```

Some fields, such as `name` or `version`, have special meanings. For instance, the `name` field is used to give the package a name. Node.js has some rules to decide what is a valid name and what is not.

For now, it is sufficient to know that valid names can be formed with lowercase letters and dashes. Since package names may appear in URLs, a package name is not allowed to contain any non-URL-safe characters.

The `version` field has to follow the specification for **semantic versioning** (**semver**). The GitHub repository at `https://github.com/npm/node-semver` contains the Node.js implementation and many examples for valid versions. Even more important is that semver also allows you to select a matching version using a range notation, which is useful for dependencies.

> **Semver**
>
> Besides the rules and constraints for version identifiers, the concept of semver is used to clearly communicate the impact of changes to package users when updating dependencies. According to semver, the three parts of a version (X.Y.Z – for example, 1.2.3) all serve a different purpose.
>
> The leading number (X) is the major version, which indicates the compatibility level. The middle number (Y) is the minor version, which indicates the feature level. Finally, the last number (Z) is the patch level, which is useful for hotfixes. Generally, patch-level changes should always be applied, while feature-level changes are optional. Compatibility-level changes should never be applied automatically, as they usually involve some refactoring.

By default, if the same directory contains an `index.js` file, then this is considered the *main*, *root*, or *entry* module of the package. Alternatively, we can specify the main module of a package using the `main` field.

To change the location of the main module of the package to an `app.js` file located within the `lib` subdirectory, we can write the following:

package.json

```
{
  "name": "my-package",
  "version": "1.0.0",
  "main": "./lib/app.js"
}
```

Furthermore, the `package.json` can be used to include some metadata about the package itself. This can be very helpful for users of the package. Sometimes, this metadata is also used in tooling – for example, to automatically open the website of the package or the issue tracker or show other packages from the same author.

Among the most useful metadata, we have the following:

- `description`: A description of the package, which will be shown on websites that list the package.

- `license`: A license using a valid **Software Package Data Exchange** (**SPDX**) identifier such as *MIT* or *BSD-2*. License expressions such as (`ISC OR GPL-3.0`) are also possible. These will be shown on websites that list the package.

- `author`: Either a simple string or an object containing information about the author (for example, `name`, `email`, or `url`). Will be shown on websites that list the package.

- `contributors`: Essentially, an array of authors or people who contributed in one way or another to the package.

- repository: An object with the url and type (for example, git) of the code repository – that is, where the source code of the package is stored and maintained.

- bugs: The URL of an issue tracker that can be used to report issues and make feature requests.

- keywords: An array of words that can be used to categorize the package. This is very useful for finding packages and is the main source of search engines.

- homepage: The URL of the package's website.

- funding: An object with the url and type (for example, patreon) of the package's financial support platform. This object is also integrated into tooling and websites showing the package.

There are a couple more fields that are necessary to specify when dealing with third-party packages. We'll cover those in *Chapter 3, Choosing a Package Manager,* when we discuss package managers in great detail.

Summary

In this chapter, you learned about a set of different module formats as alternatives to the CommonJS module format. You have been introduced to the current standard approach of writing ESMs, which brings a module system directly to the JavaScript language.

You also saw how alternative module formats such as AMD or UMD can be used to run JavaScript modules on other older JavaScript runtimes. We discussed that by using the specialized module loader, SystemJS, you can actually make use of truly convenient and current features as a web standard today. The need for import maps is particularly striking when talking about third-party dependencies.

You learned that most third-party dependencies are actually deployed in the form of packages. In this chapter, you also saw how a package.json file defines the root of a package and what kind of data may be included in package.json file.

In the next chapter, we will learn how packages using the discussed formats can be installed and managed by using special applications called package managers. We'll see how these package managers operate under the hood and how we can use them to improve our development experience.

3

Choosing a Package Manager

So far, we have learned a bit about Node.js and its internal modules. We also started to write our own modules, but we have either avoided or worked around using third-party packages.

One of the big advantages of Node.js is that using other people's code is actually quite easy. The path to doing so leads us directly to package managers. A package manager helps us to handle the life cycle of packages containing modules that can be used in Node.js.

In this chapter, we'll learn how Node.js's de facto standard package manager **npm** works. We will then go on to learn about other package managers, such as **Yarn** and **pnpm**. They all promise some advantages in terms of usability, performance, or reliability. We will take a deeper look at them to understand these advantages and who might benefit from using each of the different package managers. Finally, we'll also look at alternatives.

This chapter will help you to use third-party libraries in your code. Third-party dependencies will make you more productive and focused, and a package manager will be useful for installing and updating third-party dependencies. By the end of the chapter, you'll know the most important package managers and which one you want to pick in the context of your project.

We will cover the following key topics in this chapter:

- Using npm
- Using Yarn
- Using pnpm
- More alternatives

Technical Requirements

Some code examples for this chapter are available at `https://github.com/PacktPublishing/Modern-Frontend-Development-with-Node.js/tree/main/Chapter03`.

The CiA videos for this chapter can be accessed at `http://bit.ly/3TmZr22`.

Using npm

When you install Node.js from the official sources, you get a bit more than just Node.js. For convenience, Node.js will also add a few more programs and settings to your system. One of the most important additions is a tool called npm. Originally, npm was intended to stand for *Node.js Package Manager*, but today, it is essentially its own standalone name.

The goal of npm is to allow developers to manage third-party dependencies. This includes installing and updating packages, as well as handling their versioning and transitive dependencies. A transitive dependency is established when dependencies that are installed also include dependencies, which therefore need to be installed, too.

For npm to know what dependencies exist and what their dependencies are, the npm registry was created. It is a web service that hosts all packages on a file server.

Changing the used npm registry

Today, many npm registries exist – but only the official one located at `https://registry.npmjs.org/` is used by default. To change the registry consistently, a special file, `.npmrc`, needs to be created. If the file is created in the home directory, then the change applies to all usages. Otherwise, this file could also be created next to a `package.json` – only being applied to the designated project. Finally, to only temporarily use another registry, the `--registry` command-line flag can be used. The format of the `.npmrc` file is outlined at `https://docs.npmjs.com/cli/v8/configuring-npm/npmrc`.

To use packages from the npm registry, we'll need to use the npm command-line utility. In fact, the first thing we should do when we copy or clone the source code of a Node.js project is to run npm `install` in the directory of the project's `package.json`:

```
$ npm install
```

This will install all packages that are mentioned as runtime and development dependencies in the `package.json`. The packages are downloaded from the configured npm registry and then stored in the node_modules directory. It is good practice to avoid adding the node_modules directory to your source control. For instance, for Git, you should add node_modules to your repository's `.gitignore` file. There are several reasons for this – for example, the installation might be platform-specific or the installation may be reproducible anyway.

The npm command-line utility comes with a set of integrated commands – such as the previously shown `install` command. To see what commands are available to you, the utility can be used with the `--help` flag:

```
$ npm --help
```

```
Usage: npm <command>

where <command> is one of:
    access, adduser, audit, bin, bugs, c, cache, ci, cit,
    clean-install, [...], v, version, view, whoami


npm <command> -h   quick help on <command>
```

The --help flag also works in combination with a specific command. If you want to know which options exist for the install command, you can just type the following:

```
$ npm install --help

npm install (with no args, in package dir)
[...]
npm install <github username>/<github project>

aliases: i, isntall, add
common options: [--save-prod|--save-dev|--save-optional]
[--save-exact] [--no-save]
```

The principle of getting context-specific help is vital to many command-line utilities. All of the package managers that we'll look at in this chapter feature this approach. In the end, for us as users, this has some advantages. Instead of needing to look up the online documentation, other books, or tutorials to see the syntax for a command every time, we can just get all the required information directly in the command line which is tailored to the specific version that we use.

A command that is highly useful is init. While install is great to use for existing projects, init can be used to create a new project. When you run npm init, you'll be guided through all the options in a kind of survey. The result is shown as follows:

```
$ npm init
package name: (my-project)
version: (1.0.0)
description: This is my new project
git repository:
author: Florian Rappl
license: (ISC) MIT
About to write to /home/node/my-project/package.json:
```

```
{
  "name": "my-project",
  "version": "1.0.0",
  "description": "This is my new project",
  "keywords": [],
  "scripts": {
    "test": "echo \"Error: no test specified\" && exit 1"
  },
  "main": "index.js",
  "author": "Florian Rappl",
  "license": "MIT"
}


Is this OK? (yes) yes
```

An alternative would be to specify the -y flag. This way, all the defaults will be taken – a much quicker alternative if you just want to initialize a new project.

The initializer function of npm can even be extended. If you provide another name after npm init, then npm will try to look for a package using the create- prefix. For instance, when you run npm init react-app, npm will look for a package called create-react-app and run it. Running a package refers to looking for a bin field in the package's package.json file and using the given reference to start a new process.

If you want to add dependencies to your project instead, you can use npm install, too. For instance, adding React as a dependency is npm install react.

The dependency life cycle also requires us to know when dependencies are outdated. For this purpose, npm offers the npm outdated command:

```
$ npm outdated
Package         Current   Wanted   Latest   Location
@types/node     16.11.9   17.0.40  17.0.40  pilet-foo
react           17.0.2    17.0.2   18.1.0   pilet-foo
typescript      4.5.2     4.7.3    4.7.3    pilet-foo
```

The command only shows packages that have a more recent release than the currently installed version. In some cases, that is fine – that is, when the current version matches the wanted version. In other cases, running npm update will actually update the installed version.

> **Using different versions of npm**
>
> npm is already packaged together with Node.js. Therefore, each release of Node.js also selects a version of npm. For instance, Node.js 14 was bundled with npm 6. In Node.js 15, npm 7 was included. With Node.js 16 onward, you'll get npm 8. One way to stay flexible is to use **nvm** instead. nvm is a small tool that allows you to select the version of Node.js to use. It can also be used to change the default version and quickly update and install new versions of Node.js and npm. More information is available at `https://github.com/nvm-sh/nvm`.

npm also provides a lot of useful, convenient features – for example, to improve security. The `npm audit` command checks the currently installed packages against an online database containing security vulnerabilities. Quite often, a fix in vulnerable packages is just one call of `npm audit --fix` flag away. Furthermore, using a command such as `npm view` – for example, in `npm view react` – we can directly interact with the npm registry containing most of the publicly available packages.

While the npm registry is a great source for packages, the `npm` command-line utility is not the only way to use it. In fact, the API of the web service is public and could be used by anyone – or any program for that matter.

One of the first companies to use a public API of the npm registry was Facebook. They suffered from slow installation times in their large projects and wanted to improve on this by providing a better algorithm to actually resolve the dependencies of a project – especially transitive dependencies. The result was a new package manager named **Yarn**.

Using Yarn

The issue with the original npm package resolution algorithm was that it was created in a resilient but naïve way. This does not mean that the algorithm was simple. Rather, here, we refer to the fact that no exotic tricks or experience optimizations have been considered. Instead of trying to optimize (that is, lower) the number of packages available on the local disk, it was designed to put the packages into the same hierarchy as they were declared in. This results in a filesystem view as shown in *Figure 3.1*:

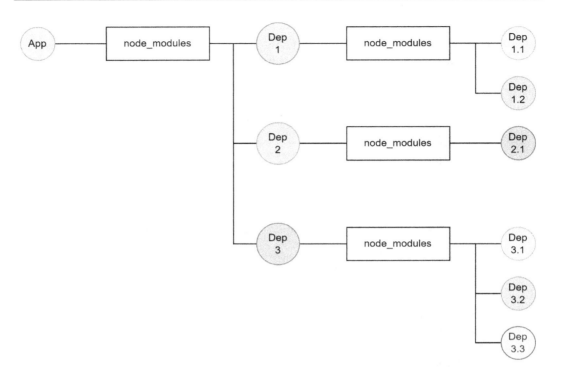

Figure 3.1 – Example filesystem snapshot after installing packages using npm

The naïve way of handling package installations is certainly a great way to ensure that everything is installed correctly, but not ideal in terms of performance. Looking at *Figure 3.1*, there may be some optimizations possible.

Let's add some example package names and versions to *Figure 3.1* to see the opportunities for optimization. In *Figure 3.2*, the same snapshot is shown – just with example package names:

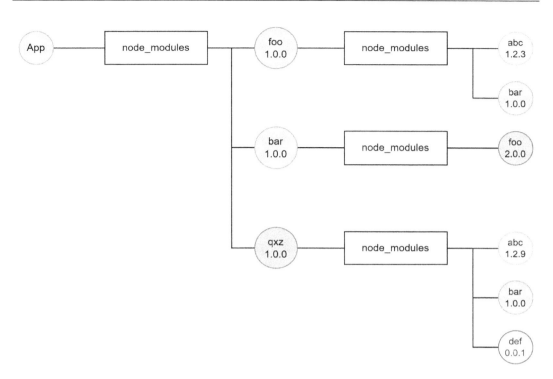

Figure 3.2 – Filesystem snapshot with example package names after npm installation

Instead of duplicating the `bar` dependency, it could be just used once. The `foo` dependency, on the other hand, has to be duplicated due to conflicting versions. Other transitive dependencies, such as `abc` or `def`, can be brought to the top level.

The resulting image is shown in *Figure 3.3*. This flattens the structure where possible. This optimization was key to the first version of Yarn. Actually, it was so successful that npm improved its algorithm, too. Today, npm resolves the packages in a similar way to the sketch shown in *Figure 3.3*:

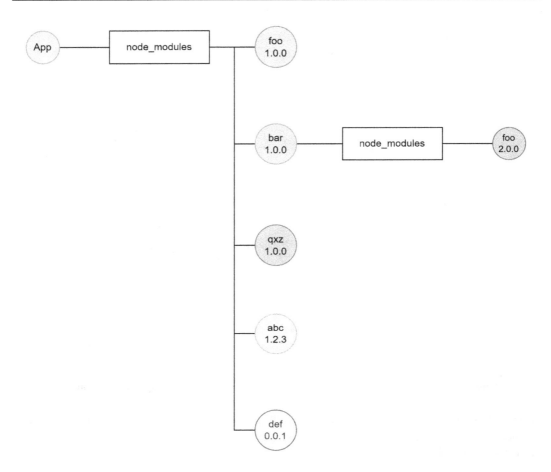

Figure 3.3 – Filesystem snapshot with example package names after installation using Yarn

For the team at Yarn, the optimizations gained were not enough. They started to look for different ways to improve even further. However, the more they looked, the more they were convinced that something completely new was needed to make any further enhancements.

The result was introduced with **Yarn 2: Plug'n'Play (PnP)**. With PnP, there is no node_modules directory. Instead, a special file called .pnp.cjs is created to give information on how the dependencies can be resolved. With the .pnp.cjs file, every package can be resolved – just as with node_ modules beforehand.

The specific location of the packages depends on the project's settings. With Yarn 2, a new concept called *zero-installs* has been introduced. This way, each dependency will be available within the project – just in a .yarn/cache subfolder. To actually achieve zero-installs, the .yarn folder should be checked into source control. Now, when the project is cloned, no installation needs to be performed. The dependencies are already part of the repository.

While most commands are very similar, Yarn takes a different approach to adding new dependencies. Here, dependencies are added using `yarn add` – for example, `yarn add react`. The installation of packages using the `yarn` command-line utility is quite similar to the previous usage with npm, though:

```
$ yarn install
▶ YN0000: ┌ Resolution step
▶ YN0000: └ Completed in 0s 634ms
▶ YN0000: ┌ Fetch step
▶ YN0013: │ js-tokens@npm:4.0.0 can't be found in the cache
and will be fetched from the remote registry
▶ YN0013: │ loose-envify@npm:1.4.0 can't be found in the cache
and will be fetched from the remote registry
▶ YN0013: │ react-dom@npm:18.1.0 can't be found in the cache
and will be fetched from the remote registry
▶ YN0013: │ react@npm:18.1.0 can't be found in the cache and
will be fetched from the remote registry
▶ YN0013: │ scheduler@npm:0.22.0 can't be found in the cache
and will be fetched from the remote registry
▶ YN0000: └ Completed
▶ YN0000: ┌ Link step
▶ YN0000: └ Completed
▶ YN0000: Done in 0s 731ms
```

In *Figure 3.4*, the new PnP mechanism is shown using the previous example. By using fully qualified names consisting of the package name and version, unique identifiers are created, allowing multiple versions of the same package to be located in a flat structure.

The downside of the PnP mechanism is the custom resolution method, which requires some patching in Node.js. The standard resolution mechanism of Node.js uses `node_modules` to actually find modules within packages. The custom resolution method teaches Node.js to use a different directory with a different structure to find modules:

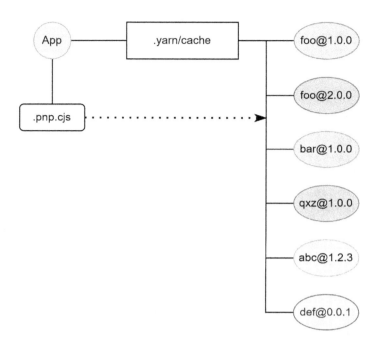

Figure 3.4 – Filesystem snapshot with example package names after installation using Yarn PnP

While using a custom resolution method is not a problem for many packages, some may depend on the classic structure involving node_modules with packages just resolving to directories and files. In PnP, however, the structure is flat, with each package being a zip archive.

As of today, many plugins and patches are available for packages to be compatible with PnP. Many – especially less popular – packages are still not usable with PnP. Luckily, Yarn 3 fixed many of these issues, providing a compatibility mode that works for most of these problematic packages. At the end of the day, it's, unfortunately, mostly a matter of trial and error. Luckily, Yarn PnP is not the only solution that exists for speeding up npm even more.

Even before Yarn 2 with PnP was released, other developers started thinking of alternative strategies to speed up installation times and preserve network bandwidth and storage capacity. The best-known attempt is a utility called pnpm.

Using pnpm

The approach of pnpm feels a bit like the original package resolution of npm. Here, each package is essentially isolated and puts its own dependencies into a local `node_modules` subfolder.

There is, however, one crucial difference: instead of having a hard copy of each dependency, the different dependencies are made available through symbolic links. The advantage of this approach is that every dependency only needs to be resolved once per system.

The other advantage is that for most packages everything is as it should be. There is nothing hiding behind an archive or via some custom mapping defined by a module that would run in the beginning. The whole package resolution just works. The exception to this rule is packages that use their path to find other packages or work against a root directory. Since the physical location of the packages is global, and therefore different from the project's location, these approaches do not work with pnpm.

Installing packages with the pnpm command-line utility works very similarly to npm:

```
$ pnpm install
Packages: +5
+++++
Packages are hard linked from the content-addressable store to
the virtual store.
   Content-addressable store is at: /home/rapplf/.local/share/
pnpm/store/v3
   Virtual store is at:             node_modules/.pnpm

dependencies:
+ react 18.1.0
+ react-dom 18.1.0

Progress: resolved 5, reused 2, downloaded 3, added 5, done
```

Overall, most commands of the pnpm command-line utility have either the same or a very similar name to their npm counterpart.

On installation, pnpm adds the unavailable packages to a local store. A local store is just a special directory from pnpm that is not bound to your project, but rather your user account. It is pnpm's package storage that is actually the source of its miraculous performance. Afterward, pnpm creates all the symbolic links to wire everything together. The result looks similar to *Figure 3.5*:

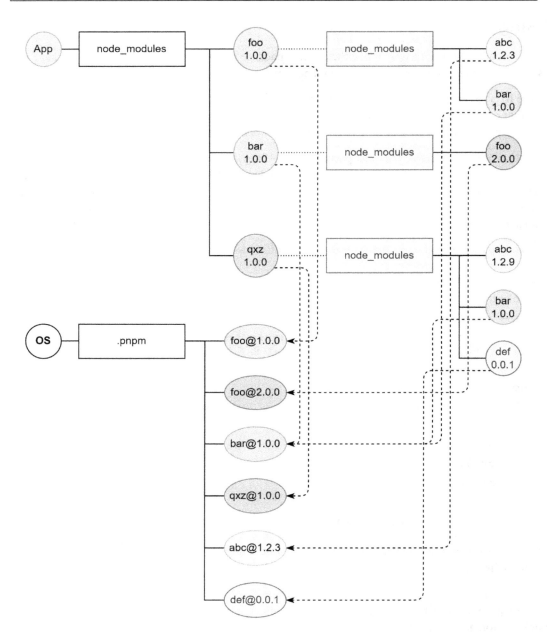

Figure 3.5 – Filesystem snapshot with example package names after installation using pnpm

Only direct dependencies are listed in the node_modules folder. The content of each subfolder is not available in the original node_modules – rather, in the global .pnpm cache. The same is then applied to all sub-dependencies.

The result is a massive performance boost. Already, on a clean install, pnpm is faster than the competition. However, in other scenarios, the relative gap may be even larger. In *Figure 3.6*, the performance of pnpm is compared against other package managers. Lower bars refer to better performance:

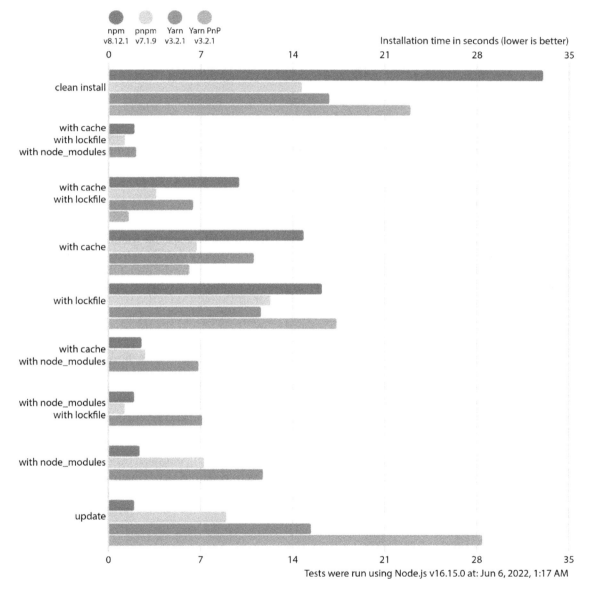

Figure 3.6 – Performance benchmark comparing pnpm against npm, Yarn, and Yarn with PnP (source: https://pnpm.io/benchmarks)

Only in the case of an up-to-date installation can npm be considered the fastest option. In other cases, pnpm and sometimes Yarn PnP can be considered faster. With this in mind, the key question is whether there are other alternatives to consider. Let's see what else we can do to simplify dependency management.

More alternatives

There is no strict requirement when using a package manager. Theoretically, it does not matter where the code comes from. You could, for instance, download the packages directly, extract them, and refer to them via their local path.

Alternatively, a system such as **Deno** could be interesting. On the surface, Deno is quite similar to Node.js. However, there are a few crucial differences under the hood. The most striking one is that there is no package manager for Deno. Instead, packages are just URLs that are resolved once needed. This way, the package installation is just a download – which happens to run when needed.

> **Deno in a nutshell**
>
> Deno was created by Ryan Dahl – the creator of Node.js. As such, Deno shares many features with Node.js but deviates in some aspects. Deno aims to be a lot more compatible with JavaScript running in the browser than Node.js. Deno also tries to be secure by default. When running a script with Deno, the provided security context has to be defined by the user. Otherwise, access to the network or the filesystem may be impossible for the code running. You can get more information at `https://deno.land/`.

Another option is to use a tool that actually leverages one of the existing package managers underneath but in a more efficient or user-friendly fashion. One example in this category is **Turborepo**.

Turborepo works with any of the popular package managers and claims to provide improved performance for many tasks, including package installation and updates. The most efficient way, however, to utilize Turborepo is to use it for a so-called monorepo, which will be discussed in greater length in *Chapter 9, Structuring Code in Monorepos*.

Besides the question of how the packages can be installed, updated, and published, the other part of package management is the package registry. In this space, you can choose from many commercial offerings to open source projects such as **Verdaccio**. Having your own package registry can be great for larger projects, where missing dependencies or downtimes of the public npm registry may be problematic.

In general, there are not many alternatives to the established package managers npm, Yarn, and pnpm. While optimizing the use of package managers or using cached registries instead might be appealing, they are certainly not worth the effort for most projects. Right now, npm and Yarn seem to be most appealing in the broadest range of scenarios, whereas pnpm could be considered the desired choice for really big repositories.

Summary

In this chapter, you learned how to use a package manager to handle everything related to packages. You've leveraged the default npm command-line utility. You got in touch with the most important alternatives, Yarn and pnpm. You should know what Yarn brings to the table – after all, PnP and zero-installs are neat features. Furthermore, you checked out some alternatives and learned about custom registries and repository task runners such as Turborepo.

At this point, you have everything to clone and run existing Node.js projects. You can install new dependencies, check for outdated dependencies, and update them. This gives you the power to integrate all of the over-a-million packages that have been published over the years in the npm registry.

In the next chapter, we will discuss how different flavors of JavaScript, such as more modern specifications or languages that use JavaScript as a compilation target, can be used in Node.js.

Part 2:
Tooling

In this part, you'll strengthen your knowledge of the Node.js ecosystem by getting in touch with a variety of tools and utilities. You'll learn how you can use different flavors of JavaScript in Node.js. Examples here include TypeScript and Flow. You'll also see which code verification and style checkers exist and how to use them.

The main focus of this part is to enable you to set up and maintain a new web development project from scratch. This also includes knowledge about quality assurance. As part of these topics, utilities such as Jest or Playwright are discussed.

This part of the book comprises the following chapters:

- *Chapter 4, Using Different Flavors of JavaScript*
- *Chapter 5, Enhancing Code Quality with Linters and Formatters*
- *Chapter 6, Building Web Apps with Bundlers*
- *Chapter 7, Improving Reliability with Testing Tools*

4
Using Different Flavors of JavaScript

With the previous chapter, you've completed the essentials for doing projects in Node.js. Looking at real projects out there, you'll find quickly that people use Node.js with all kinds of flavors of JavaScript. A **flavor** of JavaScript is a new language that can be seen as a variation of the official JavaScript language standard. Mostly, these flavors look very much like the JavaScript you are used to but differ in key parts. Sometimes, they add new language constructs to simplify certain tasks; sometimes, they bring improvements for reliability before releasing any code.

In this chapter, we'll learn how different flavors of JavaScript can be used with Node.js. We will introduce the most important tools and flavors. As far as the tooling part is concerned, we'll introduce the popular open source package, **Babel**. This tool can be quite helpful to teach Node.js how to use a flavor of JavaScript. These flavors include interesting additions to the language such as **Flow** or **TypeScript**. Both introduce type systems, but the latter also adds new constructs to the language.

This chapter will help you to use languages that can be converted to JavaScript with Node.js. Ultimately, this is key – not only to be able to run JavaScript files independent of their syntax with any version of Node.js but also to introduce additional safety and convenience in larger projects.

We will cover the following key topics in this chapter:

- Integrating Babel
- Using Flow
- Using TypeScript

Technical requirements

The complete source code for this chapter can be found at `https://github.com/ PacktPublishing/Modern-Frontend-Development-with-Node.js/tree/main/ Chapter04`.

The CiA videos for this chapter can be accessed at `http://bit.ly/3UeL4Ot`.

Integrating Babel

In the last decade, JavaScript ascended from a simple scripting language to the most used programming language in the whole world. With the increased popularity, the language has also gotten a lot of interesting features. Unfortunately, it always takes a while until the latest features are made available in all implementations. The problem gets worse if we want to use the latest language features in old implementations anyway.

This is a problem that has been known by frontend developers for years – after all, the version and variety of the browser used cannot be predetermined by the developer. Only the user makes this decision – and an older browser may not understand some of the modern features that the developer wants to use. In Node.js, we don't have exactly the same problem – as we can theoretically decide the version of Node.js – but it can be a similar issue if Node.js does not have the latest language features or if we create tools that are supposed to run on other people's machines.

A nice way out of the language feature lockdown (that is, the restriction to only use the feature set supported by the engine) is to use a tool that understands the latest language specification and is capable of properly translating it into an older language specification. The process of such a programming language translation is called **transpilation**. The tool is called a **transpiler**.

One of the most known transpilers for JavaScript is Babel. Its power lies in a rich plugin ecosystem. Actually, it is so easy to extend the JavaScript language with constructs using Babel, that many features were first introduced in Babel before they either became part of the official standard or a de facto standard. An example of the former is `async/await`, which is a fairly complex feature. An example of the latter is **JSX**, that is, the extension of JavaScript with **XML**-like constructs.

The following code is using `async/await` and would be incompatible with Node.js before version *7.6.0*:

```
function wait(time) {
  return new Promise(resolve => setTimeout(resolve, time));
}
async function example() {
  console.log('Starting...');
  await wait(1000);
```

```
    console.log('1s later...');
    await wait(1000);
    console.log('2s later...');
    await wait(3000);
    console.log('Done after 5s!');
}
example();
```

If we want to make this compatible with older versions (or, in general, JavaScript engines that cannot handle the modern `async`/`await` syntax), then we can use Babel.

There are three ways of transpiling the code with Babel:

- We can use the `@babel/node` package, which is a thin wrapper around Node.js. Essentially, it will transpile the modules during execution – that is, when they are needed.

- The `@babel/cli` package can be used to transpile the modules beforehand and run Node. js on the transpiled modules.

- Alternatively, the `@babel/core` package can be used to programmatically control the transpilation process – that is, which modules are being transpiled and what is done with the results.

Each way has its own advantages and disadvantages. For instance, choosing `@babel/node` might be the easiest to get running, but will actually give us a small performance hit and some uncertainty. If some lesser-used module has a syntax problem, then we would only find out later when the module is used.

Likewise, `@babel/cli` certainly hits the sweet spot between convenience and power. Yes, it only works with files, but that is what we want in almost all cases.

One way to see very conveniently how Babel handles things is to use the interactive website located at `https://babeljs.io/repl`. For our previous code example, which is using an `async` function with `await`, we get a view as shown in *Figure 4.1*:

Figure 4.1 – Transpiling some JavaScript via Babel online

For the screenshot shown in *Figure 4.1*, we specified the version of Node.js to be *7.6*. Once we change that to something lower, for example, *7.5*, we get a different output. It all starts with some generated code:

```
"use strict";
function asyncGeneratorStep(gen, resolve, reject, _next, _
throw, key, arg) { try { var info = gen[key](arg); var value
= info.value; } catch (error) { reject(error); return; } if
(info.done) { resolve(value); } else { Promise.resolve(value).
then(_next, _throw); } }
function _asyncToGenerator(fn) { return function () { var
self = this, args = arguments; return new Promise(function
(resolve, reject) { var gen = fn.apply(self, args); function
_next(value) { asyncGeneratorStep(gen, resolve, reject, _
next, _throw, "next", value); } function _throw(err) {
asyncGeneratorStep(gen, resolve, reject, _next, _throw,
"throw", err); } _next(undefined); }); }; }
```

After the generated code, our own code is spat out. The crucial difference is that our code now uses the helpers from the preceding generated code:

```
function wait(time) {
  return new Promise(resolve => setTimeout(resolve, time));
}

function example() {
  return _example.apply(this, arguments);
}

function _example() {
  _example = _asyncToGenerator(function* () {
    console.log('Starting...');
    yield wait(1000);
    console.log('1s later...');
    yield wait(1000);
    console.log('2s later...');
    yield wait(3000);
    console.log('Done after 5s!');
  });
  return _example.apply(this, arguments);
}

example();
```

As you can see, the code was modified with the generated functions. In our case, those functions have been used to replace the standard async/await mechanism with a generator function using yield. But even that could be changed further when transpiling for Node.js before version 6.0, which introduced support for generator functions.

In any case, Babel is actually doing the hard work of figuring out which constructs are used in our code, and which constructs need to be replaced depending on the target version of Node.js. It also knows the proper replacements and can generate some boilerplate code to support the language constructs.

For Babel to do all this work, it needs to understand the JavaScript language. This is done by **parsing** the source code. Parsing is a process that involves going over all characters, grouping them into so-called tokens (such as identifiers, numbers, etc.), and then putting these tokens in a tree-like structure known as an **abstract syntax tree** (AST). One tool to explore the AST as seen by Babel can be found at https://astexplorer.net/.

> **Understanding ASTs**
>
> Much like processing HTML results in a tree of different nodes, any programming language actually resolves to a tree of expressions and statements. While statements such as a `for` loop form a closed block of instructions, expressions such as an addition will always return a value. The AST puts all of those in relation and integrates all provided information for the respective node types. For instance, an addition expression consists of two expressions that should be added together. Those could be any expression, for example, a simple literal expression such as a number token.

A snippet of the AST of the preceding example can be seen in *Figure 4.2*. Each node in the AST has an associated type (such as `AwaitExpression`) and a position in the source document:

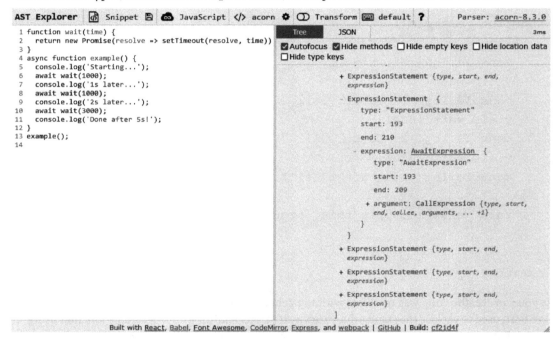

Figure 4.2 – The AST Explorer shows the information as parsed by Babel

Equipped with this knowledge, we can now try to do something locally using @babel/node:

1. We start by creating a new Node.js project. In a new directory, run the following:

   ```
   $ npm init -y
   ```

2. This will create a package.json and include some basic information. Afterwards, you can install the @babel/node and @babel/core packages as a development dependency:

   ```
   $ npm install @babel/core @babel/node --save-dev
   ```

3. Feel free to use another package manager for this. Once the installation has finished, we should add the script. Create a new `index.js` file with the following content:

index.js

```
let x = 1;
let y = 2;
// use conditional assignment - ES2021 feature
x &&= y;
console.log('Conditional assignment', x);
```

The code uses an ES2021 feature called conditional assignments. Only if `y` is truthy will the assignment be done. In this case, we expect `x` to be 2 after the conditional assignment.

4. To run the code, we need to modify `package.json`. In the `scripts` section, we add a `start` field. Now, the `scripts` section should look like this:

```
"scripts": {
  "start": "node index.js",
  "test": "echo \"Error: no test specified\" && exit
    1"
}
```

5. At this point, we can conveniently run the script using `npm start`. For recent Node.js versions (*15* or later), the output should be as follows:

```
$ npm start

> example01@1.0.0 start
> node index.js

Conditional assignment 2
```

6. However, if we try to run the code with Node.js *14*, we'll get an error:

```
$ npm start

> example01@1.0.0 start /home/node/examples/example01
> node index.js

/home/node/examples/example01/index.js:4
```

```
x &&= y;
  ^^^
```

```
SyntaxError: Unexpected token '&&='
```

Now, if you want this to run, you can switch the `start` script of `package.json` to use `babel-node` instead of the standard `node`. Trying this, however, will not work. The reason is that, by default, Babel does not understand the current environment and, therefore, cannot apply the transformations.

7. For Babel to actually understand it, we need to use the `@babel/preset-env` package. This is a preset that represents a collection of plugins. In this case, the `env` preset is a special one that intelligently looks for the right plugins depending on the current environment. Let's first install it:

```
$ npm install @babel/preset-env --save-dev
```

8. Afterward, we can integrate it by creating a new file:

.babelrc

```
{
    "presets": [["@babel/preset-env"]]
}
```

9. The file has to be placed next to `package.json`. Once there, Babel will automatically load the file and take its content as configuration input. Now the output matches our expectations:

```
$ npm start

> example01@1.0.0 start
  /home/rapplf/Code/Articles/Node.js-for-Frontend-
  Developers/Chapter04/example01
> babel-node index.js

Conditional assignment 2
```

With these instructions, you are now able to run modern code, even on older versions of Node.js. The preceding example did finally run in Node.js *14* – even though this version does not support the `&&=` new assignment operator.

There are many different packages that work together with Babel. Full languages or language extensions have been written for Babel. One of those extensions is Flow.

Using Flow

Flow is mainly a **static type checker** for JavaScript code. The purpose of a static type checker is to ensure at build time that everything works together as it should. As a result, we should see a lot fewer errors at runtime. In fact, proper usage of a static type checker will essentially eliminate all simple bugs and let us focus on solving the algorithmic and behavioral issues that would arise anyway.

In Flow, every JavaScript file can be changed to a Flow file. All that needs to be done is to introduce the @flow comment. A simple example is as follows:

```
// @flow
function square(n: number): number {
  return n * n;
}


square("2"); // Error!
```

Even though the code would work pretty well in standard JavaScript, Flow will help us by raising an error in the last line. The square function has been annotated using types for the n input argument and the return value. The colon notation separates the identifier or function head from the specified type.

Since the colon notation is not part of the JavaScript standard, we cannot just run the preceding code. Instead, we can use Babel together with the @babel/preset-flow package to strip away the Flow type annotations – keeping only the JavaScript that Node.js can understand.

Let's test this with a new project:

1. We start in a new directory by initializing an npm project and installing the necessary development dependencies:

   ```
   $ npm init -y
   $ npm install @babel/core @babel/node @babel/preset-
     flow --save-dev
   ```

2. Now, we configure Babel and change the package.json file:

.babelrc

```
{
  "presets": ["@babel/preset-flow"]
}
```

3. In `package.json`, we need to add a `start` field to the `scripts` section:

package.json

```json
{
  "name": "example02",
  "version": "1.0.0",
  "scripts": {
    "start": "babel-node index.js"
  },
  "devDependencies": {
    "@babel/core": "^7.18.5",
    "@babel/node": "^7.18.5",
    "@babel/preset-flow": "^7.17.12"
  }
}
```

Now, running `npm start` should work without any error message. However, if we run `node index.js`, then we'll actually face an error. Still, shouldn't we also face an error in this case?

4. Well, as mentioned, the Babel part is only for running. The installed preset only understands and removes the type annotations. It does not do the actual type checking. For this, we need to install another package called `flow-bin`:

```
$ npm install flow-bin --save-dev
```

5. We can run `flow` with the npx runner that comes already with npm. First, we initialize the project:

```
$ npx flow init
```

6. Then, we can type-check our solution:

```
$ npx flow
Launching Flow server for
  /home/node/examples/example02
Spawned flow server (pid=13278)
Logs will go to /tmp/flow/example02.log
Monitor logs will go to
  /tmp/flow/example02.monitor_log
Error ---------------------------------------------------------- index.js:6:8
```

```
Cannot call square with "2" bound to n because string
[1] is incompatible with number [2]. [incompatible-
call]

[2] 2| function square(n: number): number {
    3|   return n * n;
    4| }
    5|
[1] 6| square("2"); // Error!
    7|
```

```
Found 1 error
```

As expected, the call does not satisfy the type checks. This is great for our own code, but it is even better for using third-party libraries. With type checking, we can be sure that we use the provided APIs correctly. Not only now but also in the future when we install an update for third-party libraries.

Unfortunately, not every package comes with flow-type annotations. However, the situation looks a bit more promising for a quite similar tool called TypeScript.

Using TypeScript

TypeScript is a full programming language that was designed as a superset of JavaScript. The basic idea was to start with JavaScript, enhance it with missing parts such as types, classes, or enums, and choose JavaScript as a transpilation target for the language. Over the years, many of the features that were first introduced in the TypeScript language also made it to the JavaScript language.

Today, TypeScript is the most popular way to write large-scale JavaScript projects. Nearly every package on the official npm registry comes with TypeScript-compatible type annotations – either within the package or in a dedicated package. As an example, the type annotations for the react package can be found in the @types/react package.

To use TypeScript, we need to install the typescript package. This contains the tsc script, which gives us the ability to check types and transpile TypeScript files written using the .ts or .tsx extension.

Let's go ahead and create a new project, install typescript, and add a source file:

1. We start with the project creation. In a new directory, run the following:

    ```
    $ npm init -y
    $ npm install typescript --save-dev
    ```

2. Let's add an `index.ts` file with content similar to the example for Flow:

index.ts

```
function square(n: number): number {
  return n * n;
}
square("2"); // Error!
```

The content of the file is pretty much the same as beforehand, however, the `@flow` comment is missing.

3. We can now run this directly via the `tsc` command, which has been installed together with the `typescript` package:

```
$ npx tsc index.ts
index.ts:5:8 - error TS2345: Argument of type 'string'
   is not assignable to parameter of type 'number'.

5 square("2"); // Error!
        ~~~

Found 1 error in index.ts:5
```

In comparison to the `flow` tool, `tsc` does a bit more. It does not only do the type checking but it will also produce output files. What it does not do is run the code. The immediate evaluation functionality of `@babel/node` can be found in the `ts-node` package, which works quite similarly to its Babel counterpart.

4. By default, `tsc` tries to convert a `.ts` or `.tsx` input file to some new files: a `.js` and `.d.ts` file. Even in the case of failed type checks, these files might be produced. The `.js` file will be written by default, that is, with every use of `tsc`, unless we tell TypeScript to not emit the output. The `.d.ts` file will only be written if we also enable the creation of declarations. Looking at the directory after we've run the previous example will reveal two new files:

```
$ ls -l
-rw-r--r-- 1   64 index.js
-rw-r--r-- 1   79 index.ts
drwxr-xr-x 4 4096 node_modules
-rw-r--r-- 1  387 package-lock.json
-rw-r--r-- 1  278 package.json
```

5. Having the additional JavaScript is needed to actually run the code. This also applies to TypeScript being written for the browser. Since no browser understands TypeScript code, we need to transpile it to JavaScript beforehand. Like Babel, we can actually transpile for different versions of the JavaScript standard.

6. In order to keep your code repository clean, you should not use TypeScript as shown earlier. Instead, a much better way is to introduce a `tsconfig.json` file, which you should place adjacent to the `package.json`. This way, you can not only properly define the target JavaScript version but also a destination directory where the transpilation output should be placed. The destination directory can then be ignored in your version control system:

tsconfig.json

```json
{
  "compilerOptions": {
    "target": "es6",
    "outDir": "./dist"
  },
  "include": [
    "./src"
  ],
  "exclude": [
    "node_modules"
  ]
}
```

In the configuration, we indicated an `src` directory as the root for the transpilation. Every `.ts` and `.tsx` file inside will be transpiled. The output will be available in the `dist` directory.

7. Now, you can just move `index.ts` inside a new `src` subfolder and try running `tsc` again. The same error pops up, but instead of creating the `index.js` adjacent to the `index.ts` file, the output would appear in the `dist` folder:

```
$ npx tsc
src/index.ts:5:8 - error TS2345: Argument of type
  'string' is not assignable to parameter of type
  'number'.

5 square("2"); // Error!
         ~~~
```

```
Found 1 error in src/index.ts:5
$ ls -l dist/
-rw-r--r-- 1     64 index.js
```

Today, most libraries that are published on the public npm registry will be created using TypeScript. This not only prevents some unnecessary bugs but also makes the experience for consumers of the library much better.

Summary

In this chapter, you learned how to use different flavors of JavaScript with Node.js. You have seen how Babel can be installed, configured, and used to transpile your code to the JavaScript standard supported by the target version of Node.js.

Right now, you should also know the basics of the most important JavaScript flavors: Flow and TypeScript. We discussed how they can be installed and configured. Of course, to practically use these flavors, you'll need additional material to learn their syntax and master the concepts behind these languages. A good book to learn TypeScript is *Mastering TypeScript* by *Nathan Rozentals*.

In the next chapter, we will discuss a quite important area of tooling – applications that can give our code improved consistency and validation.

5

Enhancing Code Quality with Linters and Formatters

Up to this chapter, we've dealt mostly with constructs and code that has been in the hot path – that is, directly necessary to actually do something. However, in most projects, there are many parts that are not directly useful or visible. Quite often, these parts play a crucial role in keeping projects at a certain quality.

One example in the field of software project quality enhancers is the tooling that is used to ensure certain coding standards are being followed. Those tools can appear in many categories – the most prominent categories being **linters** and **formatters**. In general, these tools can be categorized as auxiliary tooling.

In this chapter, we'll learn what types of auxiliary tooling exist and why we'd potentially want to use some extra tooling to enhance our project's code quality. We'll introduce the most important auxiliary tools such as **ESLint**, **Stylelint**, and **Prettier**. We will also have a look at how these tools are integrated or used with standard text editors such as VS Code.

With the auxiliary tools presented in this chapter, you'll be able to have an outstanding positive impact on any Node.js-based frontend project that you'll contribute to.

We will cover the following key topics in this chapter:

- Understanding auxiliary tooling
- Using ESLint and alternatives
- Introducing Stylelint
- Setting up Prettier and EditorConfig

Technical requirements

The complete source code for this chapter can be found at `https://github.com/PacktPublishing/Modern-Frontend-Development-with-Node.js/tree/main/Chapter05`.

The CiA videos for this chapter can be accessed at `https://bit.ly/3fLWnyP`.

Understanding auxiliary tooling

When most people think about software, they'll have applications such as Microsoft Word, games such as Minecraft, or web applications such as Facebook in mind. Thanks to popular media, the widespread opinion is that these applications are written by individual geniuses that hack some ones and zeroes into an obscure interface. The reality could not be more far off.

As you know, to create any kind of software, lots of libraries, tooling, and – in many cases – large teams are necessary. However, what most people underestimate is the effort to just keep the ball rolling – that is, to still be able to add new features to the existing software. There are several issues that contribute to this feature slowdown.

On the one hand, the complexity within software always rises. This is whether we want it or not – with every new feature, a project becomes more challenging. In addition, larger software tends to be written by more developers – and every developer has a slightly different preference and style. This quickly becomes a mess for new developers or even those with experience in the project but who are working in areas that they did not touch beforehand.

One way to tame the rise of complexity is the introduction of processes. For instance, the process of conducting pull requests with reviews is already presented to spread knowledge about new features, detect issues, and discuss findings. At the end of a good pull request review, the code should be in a state where the new additions fit well into the whole project, both functionally and technically.

Today, everything is about automation. Therefore, while having manual processes such as a code review might be good and necessary, we usually prefer automated processes. This is exactly where all the auxiliary tooling comes in. Take, for instance, a potential discussion about code formatting within a code review. Let's say a part of the code looks as follows:

```
export function div(a,b){ return (
  a/ b)
}
```

The code itself is fine – the `div` function should perform a division, and of course, it does that. Nevertheless, the formatting is way off. A reviewer might complain that the parameters of the function should be properly formatted using a space after a comma. Another reviewer might not like the return statement, which would break without the use of parenthesis. A third review could remark on the missing optional semicolon and that the indentation is just a single space.

Now, after everything is set and done, a new version of the code would be pushed:

```
export function div(a, b){
  return a / b;
}
```

Here, the second reviewer might bring up a discussion of why the semicolon was introduced – it is only optional in this case and the code works without it. At this point, a new reviewer joins and questions the introduction of the function at all: "Why is a function for division needed in the first place? There is nothing new or interesting here."

Consequently, you'll see that much time was wasted on all sides. Instead of discussing the business need of the function in the first place, time was – and is still – spent discussing formalities that should be aligned and corrected automatically. This is where linters and formatters come into play. They can take the task of making code beautiful to read by following the standard that was set for a project. Hence, a team would need to agree only once on the tabs versus spaces debate or the use of semicolons. The tooling takes care of actually applying the decision.

Semicolons in JavaScript

JavaScript is quite loose regarding syntax. While other languages have rules and constructs that always need to be followed, JavaScript has many optional constructs in its specification. For instance, semicolons are – up to some degree – optional. There are a few cases where you'd need a semicolon to avoid nasty surprises such as in the head of `for`-loops, but for the most part, you could just drop them and your code would still work.

There are many areas in which auxiliary tooling makes sense. Sure, the alignment of code itself is nice, but even things such as commit messages when working with a project's version control system or checking whether documentation was supplied can be useful.

While checking the actual syntax – for example, the use of whitespace and newlines, is a common use case – an even more important one is to check the actual code constructs for some patterns. The validation of the used patterns is often referred to as **linting** – with a category of tools known as **linters**. A tool that excels in that space is **ESLint**.

Using ESLint and alternatives

ESLint statically analyzes code to identify common patterns and find problems. It can be used as a library from your Node.js applications, as a tool from your Node.js scripts, in your CI/CD pipelines, or implicitly within your code editor.

The general recommendation is to install ESLint locally in your Node.js project. A local installation can be done with your favorite package manager, such as npm:

```
$ npm install eslint --save-dev
```

In most cases, you'll want to specify the --save-dev flag. This will add a dependency to the development dependencies, which are not installed in consuming applications and will be skipped for production installations. Indeed, development dependencies are only interesting during the project's actual development.

Alternatively, you can also make ESLint a global tool. This way, you can run ESLint even in projects and code files that do not already include it. To install ESLint globally, you need to run the following:

```
$ npm install eslint --global
```

Potentially, you'll need elevated shell access (e.g., using sudo) to install ESLint globally. The general recommendation is to avoid using elevated shell access, which implies avoiding global installations.

Global versus local installations

npm is not only a great way to distribute packages but also to distribute tools. The standard installation of npm creates a special directory for such tools. This dedicated directory is added to your system's PATH variable, allowing direct execution of anything that is inside the directory. By using a global installation, a tool such as ESLint is added to the dedicated directory, giving us the option of running it just by typing eslint in the command line.

On the other hand, tools in a local installation are not placed in the dedicated directory. Instead, they are available in the node_modules/.bin folder. To avoid running lengthy commands such as ./node_modules/.bin/eslint, we can use the npx utility.

npx is a task runner installed together with Node.js and npm. It intelligently checks whether the provided script is installed locally or globally. If nothing is found, then a package is temporarily downloaded from the npm registry, executing the script from the temporary installation. Consequently, running npx eslint in a project where ESLint is installed will start the linting.

Let's initialize a new project (npm init -y) and install eslint as a development dependency. Now that you've installed ESLint, you can actually use it on some sample code:

1. For this, we can leverage the sample from the previous section:

---

index.js

```
export function div(a,b){ return (
  a/ b)
}
```

2. Before we can run `eslint`, we also need to create a configuration. Having a configuration file is something that almost all utilities for frontend development will require. In the case of ESLint, the configuration file should be named `.eslintrc`.

    Place the following `.eslintrc` file in the same directory as `package.json`:

.eslintrc

```
{
    "root": true,
    "parserOptions": {
        "sourceType": "module",
        "ecmaVersion": 2020
    },
    "rules": {
        "semi": ["error", "always"]
    }
}
```

There are different ways to write a configuration way for ESLint. In the preceding snippet, we used the JSON format, which should feel quite familiar for anyone with a JavaScript or web development background. Another common approach is to use the YAML format.

3. In the preceding configuration, we instruct ESLint to stop looking for parent configurations. As this is indeed the configuration for our project, we can stop at this level. Additionally, we configure ESLint's parser to actually parse ESM following a very recent specification. Finally, we configure the rule for semicolons to throw an error if semicolons are missing.

    The result of applying this ruleset can be seen in the following code snippet. Running `npx eslint` starting on all JavaScript files from the current directory (.) looks like this:

```
$ npx eslint .

/home/node/Chapter05/example01/index.js
  2:7  error  Missing semicolon  semi

✗ 1 problem (1 error, 0 warnings)
  1 error and 0 warnings potentially fixable with the
  `--fix` option.
```

As expected, the linter complains. However, this kind of complaint is certainly in the positive region. Rather constructively, ESLint also tells us about the option to automatically fix the issue.

4. Let's run the same command with the suggested `--fix` option:

```
$ npx eslint . --fix
```

No output here. Indeed, this is a good thing. The missing semicolon has been inserted:

```
export function div(a,b){ return (
  a/ b);
}
```

5. How about other rules? What if we want to force code to use anonymous arrow functions instead of the named functions? While many things can be covered by the rules coming directly with ESLint, the system can be extended with rules from third-party packages. Third-party packages that bring in additional functionality for ESLint are called ESLint plugins.

    To bring in a rule to enforce the usage of anonymous arrow functions, we can use an ESLint plugin. The package for this is called `eslint-plugin-prefer-arrow`. Let's install it first:

```
$ npm install eslint-plugin-prefer-arrow --save-dev
```

6. Now, we can change the configuration. We need to include a reference to the plugin and also specify the rule:

.eslintrc

```
{
    "root": true,
    "parserOptions": {
        "sourceType": "module",
        "ecmaVersion": 2020
    },
    "plugins": [
      "prefer-arrow"
    ],
    "rules": {
        "semi": ["error", "always"],
        "prefer-arrow/prefer-arrow-functions": ["error",
{}]
    }
}
```

7. With this configuration, we can now test whether the function declaration is indeed qualified as an error:

```
$ npx eslint .

/home/node/Chapter05/example01/index.js
  1:8  error  Use const or class constructors instead
of named functions  prefer-arrow/prefer-arrow-
functions

✗ 1 problem (1 error, 0 warnings)
```

In contrast to the previous error, we are not seeing any hint of an automatic fix here. In such cases, the author of the code has to do all the changes to please the linter manually.

There are plenty of alternatives to ESLint. In the past, the TypeScript-specific variant TSLint was quite popular. However, a couple of years ago, the team behind TSLint decided to actually merge their rules into ESLint – also making ESLint the de facto standard for linting TypeScript files. Today, the most popular alternatives are **Rome**, **quick-lint-js**, and **JSHint**.

Rome is an all-in-one tooling that combines several utilities into one cohesive application. While Rome is not written in JavaScript using Node.js, it still integrates nicely into the standard frontend tooling. One of the aspects covered in Rome is linting. At the time of writing, Rome is, unfortunately, not yet feature-complete and still in an early alpha version, but its performance and convenience benefits are a clear advantage.

The **quick-lint-js** package is a small tool that does not require configuration and is tuned to outperform ESLint in execution time. The downside of this is that quick-lint-js has fewer features and is less flexible in its design.

Lastly, one of the golden classics in the linting field is **JSHint**. Originally, it was created as a more configurable version of **JSLint**, which can be considered the first popular linter for JavaScript. One of the issues with JSHint is that it does not support the latest and greatest features of the **ECMAScript** standard. If you are looking for ES2020 support, then JSHint can be discarded. Likewise, JSHint is a bit more restrictive on extensibility. In JSHint, you cannot define custom rules. However, if something is missing in JSHint, you'll not be able to just add it.

The biggest advantage of ESLint, however, is that it already has the ecosystem that others are potentially missing. One area where ESLint shines is editor support. *Figure 5.1* shows the entry of the official ESLint extension on the VS Code Marketplace.

Figure 5.1 – Entry of the official ESLint extension on the VS Code Marketplace

Similar plugins exist for other editors. Some editors such as Brackets even come with an ESLint integration preinstalled.

The editor integration will indicate ESLint issues directly in the code. This is really helpful during development. Instead of waiting for code quality inspection results after the code has been written, you can directly see issues when they appear. This way, you'll be able to fix them immediately, instead of being required to come back to previously closed files later.

In almost all editor integrations, you'll not only get some squiggles or similar visual hints when ESLint finds an issue but also the possibility to run a quick fix. Running a quick fix will trigger ESLint's repair functionality. In the preceding command line usage, we triggered this behavior by using the --fix flag.

Figure 5.2 shows how VS Code reports the issues found by ESLint on the given example file, index.js:

Figure 5.2 – ESLint integration reporting issues in VS Code

In general, it makes sense to define a sound basis for linting rules. The basis, however, should not be too large. Too many rules will ultimately have the opposite effect. Instead of empowering a team by finding a common style and avoiding problematic patterns, the constraints are too heavy – essentially slowing down or even blocking progress on new features. Therefore, the recommendation is to start with a few rules and add new ones when certain code issues come up more often in pull request reviews. This way, the set of linting rules will evolve with the project.

While linting JavaScript source files is definitely one of the most important tasks, it is by far not the only kind of source file you'll encounter in modern frontend development. Arguably, the second most important type of files are stylesheets such as **CSS** or **SCSS**. For those, we can rely on another tool for linting called Stylelint.

Introducing Stylelint

Stylelint is a linter for CSS files and can be extended to also understand CSS dialects such as SCSS, **Sass**, **Less**, or **SugarCSS**. It has over 170 built-in rules but, much like ESLint, provides support for custom rules.

To install Stylelint, we can follow the same steps as with ESLint:

1. Here, it usually makes sense to rely on the standard configuration provided by Stylelint. Unlike ESLint, the standard configuration is released in a separate package and, therefore, needs to be installed as well. The command to install both packages as development dependencies looks like this:

    ```
    $ npm install stylelint stylelint-config-standard
      --save-dev
    ```

2. In any case, we still require a configuration file. For the moment, it is sufficient to just let stylelint know that we want to use the configuration from the stylelint-config-standard package. Here, we can write another configuration file next to the project's package.json:

.stylelintrc

```
{
    "extends": "stylelint-config-standard"
}
```

3. Next, let's introduce some problematic CSS files to try out the `stylelint` utility:

style.css

```
div {
    padding-left: 20px;
    padding: 10px;
}
p {
    color: #44;
}
```

The preceding snippet has several issues. On one hand, we'll overwrite the `padding-left` property by using the `padding` shorthand afterward. On the other hand, we'll use an invalid color hex code. Finally, we might want to have a new line between different declaration blocks.

4. We can run the `stylelint` utility with the npx task runner – just like how we triggered `eslint`:

```
$ npx stylelint style.css

style.css
  2:5   ✘   Expected indentation of 2 spaces
        indentation
  3:5   ✘   Unexpected shorthand "padding" after
        "padding-left"  declaration-block-no-shorthand-
        property-overrides
  3:5   ✘   Expected indentation of 2 spaces
        indentation
  5:1   ✘   Expected empty line before rule
        rule-empty-line-before
  6:5   ✘   Expected indentation of 2 spaces
        indentation
  6:12  ✘   Unexpected invalid hex color "#44"
        color-no-invalid-hex
  7:1   ✘   Unexpected missing end-of-source newline
        no-missing-end-of-source-newline
```

5.   There is quite a list of issues coming out! Luckily, just like with `eslint`, we can use the `--fix` flag to automatically fix as much as possible:

```
$ npx stylelint style.css --fix

style.css
  3:5   ✖   Unexpected shorthand "padding" after
   "padding-left"  declaration-block-no-shorthand-
   property-overrides
  6:12  ✖   Unexpected invalid hex color "#44"
   color-no-invalid-hex
```

While the cosmetic issues dealing with spaces and newlines can be tackled automatically by Stylelint, the remaining two issues (3 : 5 and 6 : 12) require a bit more brainpower to be fixed. The first issue requires a decision of whether we want to either remove the `padding-left` property or move it after the usage of the `padding` shorthand. The second issue requires us to actually think about a valid color to use. Here, Stylelint cannot possibly know which color we had in mind when writing the code.

Stylelint is not only very useful but is also quite unique. In the world of CSS linters, there are not many options. Mostly, people tend to rely on their tooling – for example, Sass or Less, to already give them some errors and warnings. Stylelint goes a bit further. In addition to the rich set of in-built rules and its flexibility via plugins, Stylelint also offers a rich ecosystem. As with ESLint, many editors have an integration for Stylelint.

With all the linting in place, we can now turn to an even softer part of code cosmetics – how code is structured visually. A tool to help us here is **Prettier**.

Setting up Prettier and EditorConfig

Prettier is a code formatter that works with a lot of different source files. Among the supported file types, we have plain JavaScript, Flow, TypeScript, HTML, CSS, SASS, Markdown, and many more. Prettier is also integrated into many different editors such as Atom, Emacs, Sublime Text, Vim, Visual Studio, or VS Code.

Let's dig into installing and configuring the Prettier formatter:

1.   Such as the previous tools, Prettier can be installed locally or globally. Adding Prettier to an existing project can be done by installing the `prettier` package from the npm registry:

```
$ npm install prettier --save-dev
```

2. Prettier can format JavaScript code even without any configuration. To run Prettier on an existing code file, you can use the `prettier` utility with npx. For instance, to apply formatting to your previous code file, you can run:

```
$ npx prettier index.js
export function div(a, b) {
    return a / b;
}
```

In this case, Prettier just printed the result of the formatting in the command line. It also added a semicolon to the end of the statement. Let's configure Prettier to *not* add semicolons at the end of statements.

3. To configure Prettier, a `.prettierrc` file should be added to the root of the project – right next to `package.json`. The file can be written with JSON. An example is shown here:

.prettierrc

```
{
    "tabWidth": 4,
    "semi": false,
    "singleQuote": true
}
```

The provided example sets the indentation to four spaces. It instructs Prettier to always use single quotes instead of double quotes for strings when possible. Most importantly, we disable the use of semicolons.

4. With the preceding configuration in place, we can run `prettier` again:

```
$ npx prettier index.js
export function div(a, b) {
    return a / b
}
```

The effect is striking. Now, four spaces instead of two are being used. The semicolon is dropped. The configuration has been successfully applied. However, one thing that is still left open is to actually overwrite the existing file. After all, getting the formatting code in the command line is nice but not worth a lot if we did not really format the original file.

5. For `prettier` to apply the changes, the `--write` flag needs to be used. The command from *step 4* would therefore change to the following:

```
$ npx prettier index.js --write
index.js 41ms
```

The output now prints a summary of all the files that have and have not been changed. With the preceding command, only the `index.js` file is formatted; however, the `prettier` utility would also accept wild cards such as `*` to indicate placeholders matching multiple files.

Globs

Many Node.js utilities accept a special kind of syntax to match multiple files. Very often, this syntax comes directly or is at least inspired by the `glob` package, which copied the notation from Unix. The syntax defines so-called globs – that is, patterns that allow matching files. In this regular expression-like syntax, `*` matches 0 or more characters in a single path segment, while `?` matches exactly a single character. Another useful construct is `**`, which can be used to denote 0 or more directories. A pattern such as `**/*.js` would thus match any `.js` file in the current directory and any subdirectory. More details on the `glob` package and its syntax can be found at `https://www.npmjs.com/package/glob`.

While Prettier is great for many kinds of source files, it surely cannot tackle text files in general. Quite often, however, we want to establish general formatting rules for anything in our project. This is where **EditorConfig** comes in.

EditorConfig is a standard to help maintain consistent coding styles for a project. It is established by a file named `.editorconfig`. Pretty much every editor supports this file.

An `.editorconfig` example looks like the following:

.editorconfig

```
root = true
[*]
end_of_line = lf
insert_final_newline = true
indent_style = space
indent_size = 2
```

As with ESLint, we can use nested configuration files – that is, specialize the configuration for subdirectories by having another `.editorconfig` file in them. The `root = true` configuration tells the editor to stop traversing the file system upward for additional configuration files. Otherwise, this file has only a single section, `[*]`, matching all text files.

The ruleset in the preceding example above would actually tell an editor to exclusively use the line feed (`lf`) character to end lines. While this is the standard on Unix-based systems, Windows users would usually get two characters to end lines: line feed (`lf`) and carriage return (`cr`) – the so-called `lfcr` convention. In addition, the ruleset would introduce an empty line at the end of each file. By definition, each text file would use two spaces as an indentation level.

While such a configuration is nice, it can be in direct conflict with the Prettier configuration. However, another great thing about Prettier is that it can work hand in hand with EditorConfig. Let's rewire the previous configuration to also use EditorConfig:

.prettierrc

```
{
  "semi": false,
  "singleQuote": true
}
```

Since Prettier rules will always take precedence and overwrite the ones from the `.editorconfig` file, it makes sense to remove conflicting rules. Otherwise, we will be only left with the JavaScript-specific formatting rules – for example, for semicolons and the preferred quote style, in `.prettierrc`. The general text formatting rules are now specified via EditorConfig implicitly.

With all this in mind, let's recap what we've learned in this chapter.

Summary

In this chapter, you learned how code quality can be enhanced with the help of linters and formatters. You can now use common tools such as EditorConfig, Prettier, Stylelint, or ESLint. You are now able to add, configure, and run these tools in any project that you like.

At this point, you can contribute to pretty much all frontend projects that are based on Node.js for their tooling. Also, you can introduce great quality enhancers such as Prettier. Once successfully introduced, these tools ensure that certain quality gates are always fulfilled. In the case of Prettier, discussions about code style are mostly a thing of the past – helping teams all around the world to actually focus on the actual problem instead of dealing with code cosmetics.

A downside to keep in mind is that most of these tools have some assumptions about your code. So, if your code uses, for instance, one of the flavors we discussed in *Chapter 4, Using Different Flavors of JavaScript*, then you'll most likely need to teach some of your tools about this flavor, too. Quite often, this only requires the installation of a plugin, but in severe cases, you are left with the decision to either abandon the tool or stop using the flavor for your project.

In the next chapter, we will take an in-depth look at perhaps the most important tooling for frontend developers: bundlers.

6

Building Web
Apps with Bundlers

In the previous chapter, we covered an important set of auxiliary tooling – linters and formatters. While code quality is important, the undoubtedly most important aspect of every project is what is shipped and used by the customer. This is the area where a special kind of tooling – called bundlers – shines.

A bundler is a tool that understands and processes source code to produce files that can be placed on a web server and are ready to be consumed by web browsers. It takes HTML, CSS, JavaScript, and related files into consideration to make them more efficient and readable. In this process, a bundler would merge, split, minify, and even translate code from one standard such as ES2020 into another standard such as ES5.

Today, bundlers are no longer nice to have, but necessarily used for most projects directly or indirectly. Pretty much every web framework offers tooling that is built upon a bundler. Often, the challenge is to configure a bundler so that it understands our code base and does exactly what we'd expect it to do. Since web code bases are quite different, bundlers need to be flexible in many directions.

In this chapter, you'll build up an understanding of what bundlers do and how you can control their inner processes. We'll also introduce the most important bundlers as of today, and see how they can be used and configured to work efficiently for us. This will help you get your web project from raw source code to production-ready artifacts.

We will cover the following key topics in this chapter:

- Understanding bundlers
- Comparing the available bundlers
- Using Webpack
- Using esbuild
- Using Parcel
- Using Vite

Technical requirements

The complete source code for this chapter is available at `https://github.com/PacktPublishing/Modern-Frontend-Development-with-Node.js/tree/main/Chapter06`.

The CiA videos for this chapter can be accessed at `https://bit.ly/3G0NiMX`.

Understanding bundlers

Writing a modern web application is quite difficult. One of the reasons for the level of difficulty is the large variety of different technologies that need to be involved in the process. Let's just mention a few:

- HTML for writing documents
- CSS for styling those documents
- JavaScript with the DOM API to bring some interactivity
- A JavaScript UI framework to create interactive components
- A CSS preprocessor to use variables, nesting, and more features for CSS
- Potentially TypeScript or some other typed system to improve reliability in certain source code areas
- Service and web workers need to be mastered
- All static files should be easy to cache

Before the introduction of a new class of tooling that was capable of building up module graphs, dedicated task runners such as **Grunt** or **Gulp** were used. These runners were inspired by more generic approaches such as **Makefiles**. The problem, however, was that two aspects – the build process and the source code – needed to be kept in sync. Just adding one file to the source code was not sufficient; the build process had to be informed about this new file. With bundlers, this all changed.

At its core, a bundler is a tool that leverages other tools. The most important addition is that a bundler understands the module graph – that is, the relationships (imports and exports) of code modules such as the CommonJS or ESM modules we discussed in the previous chapters. It can build up a module graph and use that to let other tools such as Babel work.

To get started, a bundler requires so-called entry points – quite often, these are referred to as entries. These are files that are used as roots in the module graph. These files may depend on other files, in which case the bundler will continue in these files to build up the module graph.

Figure 6.1 shows an example module graph constructed from two entry points. The interesting property of this graph is that the content of **Entry 2** is fully contained in **Entry 1**, too. In many situations, there won't be any significant overlap between the module graphs of multiple entry points:

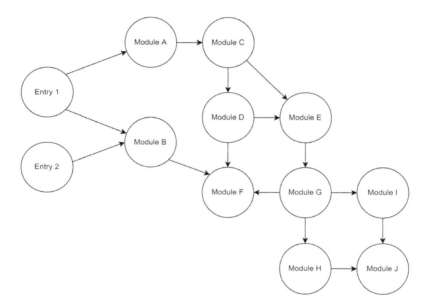

Figure 6.1 – Example module graph constructed from two entry points

Most bundlers work in phases. While each bundler uses slightly different terminology, they almost always come with the following high-level phases:

1. Module resolution
2. Module transformation
3. Chunk and asset generation
4. Applying optimizations

The module transformation is often necessary. On the one hand, the bundler needs to understand the module to find imported modules to build up the module graph; on the other hand, the chunk generation must rely on normalized input modules.

While the transformation phase needs to work hand-in-hand with a resolver to continuously build up the module graph, all other phases are pretty much independent. Quite often, the optimization phase is either reduced or fully disabled during development. This reduction helps speed up the bundling process by a fair margin. Additionally, further instructions that are quite helpful during debugging will be kept.

Minification

One of the most common optimizations is minification. The goal of minification is to make the files as small as possible without using active compression. While minification on the surface is rather easy and efficient in a language such as JavaScript, other languages such as CSS or HTML are a bit more problematic. Especially minification of HTML has been proven to be a hard problem without as many gains compared to the minification of JavaScript. After minification, files are usually not as readable as they were previously. One reason is the removal of unnecessary whitespace, which was introduced to give the original code readability and structure.

The whole bundling process can be sketched in a diagram. *Figure 6.2* shows how the different entries enter the different phases:

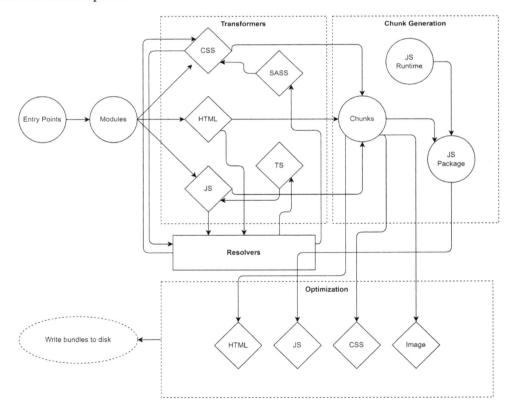

Figure 6.2 – High-level phases of a modern web bundler

Another thing to consider is that the chunk generation will mostly also introduce some kind of JavaScript runtime. This can be as lightweight as teaching the resulting code how to load additional bundles that have been created as script files, but it can also include full support for loading shared dependencies from foreign code and more. The introduced code is fully bundler-specific.

With that in mind, let's see what kind of bundlers are out there and how they compare.

Comparing available bundlers

There are multiple generations of bundlers. The first generation was centered around the belief that Node.js applications are the only kind of applications that should be written. Therefore, changing these applications into JavaScript files that work in the browser has been the primary concern of the bundlers from that generation. The most popular one in that category is **Browserify**.

The second generation went on to extend the idea from the first generation to pretty much all JavaScript code. Here, even HTML and CSS assets could be understood. For instance, using @import rules in CSS would extend the module graph to another CSS module. Importantly, while the *CommonJS* (or later on, *ESM*) syntax was still used to derive the JavaScript module graph, these second-generation bundlers did not care about Node.js. They always assumed that the code was written for the browser. Quite often, however, you could change the target and also bundle code for Node.js with them. The most popular one in this category is **Webpack**, even though Webpack always tried to go with the times and adapt.

Third-generation bundlers introduced a much-improved user experience. They tried to find a native or obvious way of handling things and often did not even require any configuration. The most popular tool in this category is the original *Parcel* bundler.

The current fourth-generation bundlers are all about performance. They either come with a dedicated runtime or sit on top of natively written tooling, which usually outperforms their older JavaScript-written counterparts. Here, we see tools such as **esbuild** or experimental runtimes such as **Bun**.

The big question is: When should you use what? With half a dozen popular bundlers and more available, the question is not easy to answer. Surely, if a team is already really familiar with one of the options, then going with it for a project is very often the right choice. Otherwise, look for similar projects and try to understand what bundler they picked – and why. In any other case, you could use the following catalog of questions to identify which bundler might be the best option:

- What kind of assets are involved? If only JavaScript is involved, then Webpack might be good. If you have multiple HTML pages that all need to be processed, then **Vite** might be a great option.

- How many dependencies are you using? Especially when you use older libraries from npm, a bundler with a broad range of support – such as Webpack – may be the best choice. Otherwise, look for faster options, such as esbuild.

- How familiar is the team with bundlers and their options? If the team is not familiar with bundling at all, then Parcel could be a great way to get started. Otherwise, Webpack potentially has the most documentation out there. A community that is rather new and very active and helpful can be found with Vite.

- Are you building an application or just want to optimize the assets of a library? Especially for a library, something smaller, such as esbuild, might be useful. On the other hand, Parcel has a lot to offer here, too. In general, Vite should be avoided for libraries. Support is there, but it just does not feel to be ready yet for building libraries more efficiently than Rollup.js and esbuild.

- Do you need support for advanced scenarios such as offline mode or web workers? In these cases, the ecosystem of Webpack is very often hard to beat. Parcel also does a good job of offering helpers in this area. esbuild should be avoided for such scenarios.

- How important is performance? If you have a larger code base (above 1,000 modules or 100,000 lines of code), then Webpack is known to be a performance killer, taking easily 30 seconds to minutes. Picking something such as Vite or – if possible – esbuild will certainly help speed up the process. While the former is more developer friendly, it also comes with a lot of hidden complexity. The latter is more direct but lacks standard features such as **hot-module reloading (HMR)**.

- How much maintenance is acceptable? Bundlers that rely on a lot of plugins are traditionally much harder to maintain. Upgrading Webpack to the next major version has been notoriously difficult. From missing plugins to breaking changes in the plugin's API – everything that can happen will also happen in such cases. Prefer bundlers such as Parcel or Vite that try to come with everything necessary out of the box.

- How important are additional development features such as bundle insights? If these are supercritical, then nothing is better than Webpack. As the Webpack ecosystem is super large, you'll find additional tools, libraries, and guides easily. On the other hand, choosing something with a growing community such as Vite might also work fine. If something is missing, the community should be capable of picking it up quickly, too.

In the following sections, we'll go over an example project to see how some of the most popular bundlers can be used to build it. We'll use a project with a small, but non-trivial code base. For this example, we'll use **React** – but don't worry, you don't need to know React to follow this chapter.

> **React**
>
> React is arguably the most popular UI library for web frontend development. It allows developers to build UIs quickly in JavaScript by leveraging a language extension known as **JSX**. By using JSX, we can write code that looks quite similar to HTML but is transpiled to JavaScript function calls. In React, the basic building block of the UI is a component, which is very often just a plain JavaScript function. By convention, the names of components usually start with an uppercase letter; for example, `Button` or `Carousel`.

The code base for the example we'll cover consists of the following:

- The source code of a **single-page application (SPA)**
- An **HTML** file as the entry point (`index.html`) of the SPA
- Several asset files (videos, images in different formats, audio)
- Several non-trivial dependencies
- Some files that use TypeScript instead of JavaScript
- A special **CSS** preprocessor called **SASS**
- A web framework (React with React Router) is being used
- Different virtual routes should lead to different pieces of the page that have to be lazy loaded

All in all, this example should produce a small demo application that contains a video and audio player that uses third-party dependencies.

Lazy loading

Lazy loading describes a technique where not all parts required by an application are loaded immediately. For a SPA, this makes sense – after all, not every component or part of the SPA will be required for the current user interaction. And even if it isn't required, it could be at some later point in time. Lazy loading usually involves loading additional script (or other) files when some action such as a user clicking on a button or following some internal link is performed. The implementation of lazy loading needs to be supported by the respective UI framework (for example, React has a function called `lazy`) but is done by the bundler.

The boilerplate for this example can be created by initializing a new Node.js project:

```
$ npm init -y
```

Now, we'll add all the runtime dependencies – that is, the packages that will be required when our application runs in the browser:

```
$ npm i react react-dom react-router-dom video.js --save
```

At the end of the day, it will be the job of the bundler to add the preceding dependencies to scripts that can be run in the browser. However, for us, it makes sense to do this to get a clear view of which packages are just required for the tooling, and which dependencies are needed for the code to run.

The basic `devDependencies` – that is, the ones that are required for the tooling – are as follows:

```
$ npm i typescript sass @types/react @types/react-dom --save-
dev
```

Additional tooling dependencies are required, too, but will be bundler-specific.

The example application contains the following source files:

- `index.html`: Template for the SPA website
- `script.tsx`: Starts to run the application
- `App.tsx`: The application root
- `Layout.tsx`: The layout of the application
- `Home.tsx`: The home page containing links to all pages
- `Video.tsx`: The page containing the video player
- `Audio.tsx`: The page containing the audio player
- `Player.jsx`: The React component for the video and audio player
- `earth.mp4`: Video file to play
- `river.webp`: Preview image (`.webp` format) for the video file
- `snow.jpg`: Preview image (`.jpg` format) for the sound file
- `sound.mp3`: Audio file to play

The process of showing a UI is usually called rendering. When React first renders the application, it needs to mount its component tree on the DOM tree. This is done in the `script.tsx` file:

script.tsx

```
import * as React from 'react';
import { createRoot } from 'react-dom/client';
import './style.scss';
import App from './App';

const root = createRoot(document.querySelector('#app')!);
root.render(<App />);
```

The usage of angle brackets for initiating `App` is referred to as JSX. Under the hood, the additional `x` in the file extension (`tsx`) enables such expressions to be processed, where `<App />` will be transformed into `React.createElement(App)`.

The App component itself is defined as follows:

App.tsx

```
import * as React from "react";
import { BrowserRouter, Route, Routes } from
  "react-router-dom";
import Layout from "./Layout";

const Home = React.lazy(() => import("./Home"));
const Video = React.lazy(() => import("./Video"));
const Audio = React.lazy(() => import("./Audio"));

const App = () => (
  <BrowserRouter>
    <Routes>
      <Route path="/" element={<Layout />}>
        <Route index element={<Home />} />
        <Route path="video" element={<Video />} />
        <Route path="audio" element={<Audio />} />
      </Route>
    </Routes>
  </BrowserRouter>
);

export default App;
```

This kind of structure is typical for a SPA. All the different routes come together in a router or root component to be displayed when a certain path is found. For instance, in our application, the /video path would show the Video component, while the /audio path would show the Audio component. All these components will be embedded in a Layout component, which is responsible for the general layout, such as showing the header and the footer, of the application.

In the App.tsx file, lazy loading is initiated by using the ESM import function. Bundlers should be capable of transforming that into loading another script and returning a Promise at that location.

Promises

The specification describes an `import` function to return a `Promise`. A `Promise` is an object that can be used to determine when an asynchronous operation is finished. The object exposes functions, which are called with the result of the asynchronous operation or with an error that was thrown during the operation. The most important functions are `then` and `catch`. The former can be used to define what to do when something is successfully returned, while the latter can be used to handle errors.

In a SPA, it makes sense to put every page in a router into lazy loading. *Figure 6.3* shows a high-level overview of the example application's modules. The dashed boxes indicate bundling areas – that is, source files that can be grouped into combined output files. This bundling is one of the most crucial aspects of any bundler:

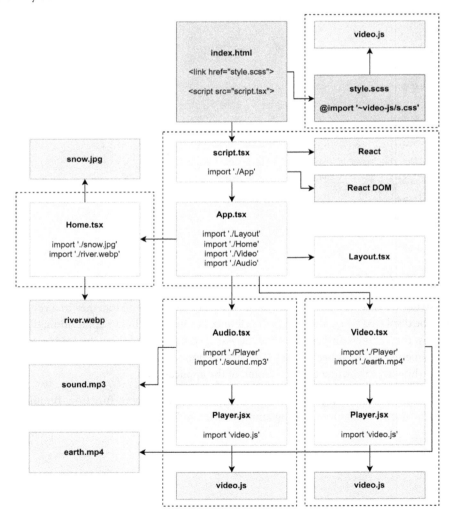

Figure 6.3 – The example application's modules

While some of the given aspects should be rather simple to implement in a bundler, other properties of the example application might be challenging. For instance, what is the behavior of a bundler when duplicate modules are found? Some bundlers may duplicate the generated code while others may create a shared bundle that is a loading prerequisite for the generated side bundles.

In the case of this example, we can see that `Player.jsx` appears twice. We'll use this to answer the question for each bundler. Furthermore, pretty much every module requires `react`; however, since it is already required in the initial script module (`script.tsx`), it should not be duplicated.

Without further ado, let's see how this example application can be bundled using Webpack.

Using Webpack

Webpack is presumably the most popular option among the available bundlers. It is also among the oldest bundlers – dating back to a time when Node.js was still young and the whole idea of bundling was rather new. At this time, task runners were still dominantly used. However, the increasing complexity of frontend development opened the door for much more elaborate tooling.

One thing that makes Webpack stand out is its ecosystem. From the very beginning, Webpack decided to develop only a very shallow core focusing on module resolution. In some sense, Webpack is just the wrapper holding all these plugins together with a fixed plan of execution. It pretty much combines the configuration that was thrown in by the user, with the power of all the selected plugins.

Today, Webpack can also work without plugins or a configuration. At least in theory. In practice, every project that goes beyond some simple examples will require a bit of configuration. Also, interesting features such as support for other languages such as TypeScript will require a plugin.

To get started with Webpack, we need to install the `webpack` and `webpack-cli` packages using npm:

```
$ npm install webpack webpack-cli --save-dev
```

If we only wanted to use Webpack programmatically, such as from a Node.js script, then we could also spare the `webpack-cli` package installation.

To run Webpack from the command line, you can use `npx` together with the `webpack` executable:

```
$ npx webpack
```

Just running Webpack like this will not be successful:

```
assets by status 0 bytes [cached] 1 asset


WARNING in configuration
The 'mode' option has not been set, webpack will fallback to
'production' for this value.
```

```
Set 'mode' option to 'development' or 'production' to enable
defaults for each environment.
You can also set it to 'none' to disable any default behavior.
Learn more: https://webpack.js.org/configuration/mode/

ERROR in main
Module not found: Error: Can't resolve './src' in '/home/node/
examples/Chapter06/example01'
[...]
webpack 5.74.0 compiled with 1 error and 1 warning in 116 ms
```

Fixing the warning about mode is rather simple – all we need to do is to provide a CLI flag such as
--mode production. The more problematic issue is that Webpack does not find any entry point.

As mentioned already, there is a chance that Webpack may just work, but usually, we'll be forced to
create a configuration file. Webpack uses real Node.js modules to provide the configuration, which
gives us the full power of the Node.js ecosystem. A Webpack configuration is called webpack.
config.js and should be placed next to the package.json file.

Let's create a rather lightweight one. The highlighted property is one of Webpack's fundamental
configuration sections, telling the bundler what entry points to use:

webpack.config.js

```
module.exports = {
  entry: {
    app: "./src/script.tsx",
  },
};
```

Now, we can try running Webpack again:

```
$ npx webpack --mode production
assets by status 360 bytes [cached] 1 asset
./src/script.tsx 185 bytes [built] [code generated] [1 error]

ERROR in ./src/script.tsx 5:54
Module parse failed: Unexpected token (5:54)
You may need an appropriate loader to handle this file type,
currently no loaders are configured to process this file. See
https://webpack.js.org/concepts#loaders
```

```
|  import App from './App';
|
>  const root = createRoot(document.querySelector('#app')!);
|  root.render(<App />);
|
```

```
webpack 5.74.0 compiled with 1 error in 145 ms
```

This is better, but we're still not there yet. Webpack requires a plugin to understand special files such as TypeScript or SASS sources. Therefore, we need to install these development dependencies, too. In this case, we require quite a list of plugins to make everything work:

- `ts-loader` is a plugin for handling TypeScript files by transforming them into JavaScript
- `sass-loader` is a plugin for handling SASS files by transforming them into CSS
- `css-loader` is a plugin for handling CSS by transforming it into a text module
- `style-loader` is a plugin for handling CSS by transforming it into a JavaScript module
- `babel-loader` is a plugin for using Babel to transform JavaScript files with additional syntax (such as JSX) into plain JS
- `html-webpack-plugin` is a plugin for loading an HTML file as a template for the output HTML file

The big disadvantage of Webpack is that everything must be a JavaScript module in the end. Quite often, plugins perform some tricks to end up with empty modules, but they still emit the result (such as a separate image or CSS file) to the filesystem.

You can install the remaining dependencies can be done from the command line:

```
$ npm i ts-loader sass-loader css-loader style-loader babel-
  loader @babel/core @babel/preset-env @babel/preset-react html-
  webpack-plugin --save-dev
```

One thing we also need to supply is a proper `tsconfig.json`. Without this file, TypeScript won't be configured correctly. The `ts-loader` plugin of Webpack works quite closely together with TypeScripts, so it requires this file to know what files to consider and which files to drop. It also uses this to properly transform the file:

tsconfig.json

```
{
  "compilerOptions": {
```

```
      "jsx": "react",
      "module": "ESNext"
   },
   "include": ["./src"],
   "exclude": ["./node_modules"]
}
```

In this configuration, TypeScript has been set up to process JSX in the default React way (that is, transforming JSX into `React.createElement` calls). The configuration will also output ESM module syntax (the highlighted option), which is important for Webpack to correctly identify imports and exports. Without this, bundle splitting won't work if triggered from TypeScript files. Finally, we include all the files in the `src` folder and exclude the `node_modules` folder. The latter is a common practice to save a substantial amount of processing time.

Now, to get all these things working together, we need to extend the Webpack configuration quite a bit. First, we need to import (that is, `require`) all the plugins that we'd like to use. In this case, we only want to use `html-webpack-plugin`. Next, we need to set up the rules for all the loaders we need to include. This is done via the `module.rules` array. Finally, we need to define what to do with the remaining assets.

Let's see how the Webpack configuration could be written to successfully bundle our example:

webpack.config.js

```js
const HtmlWebpackPlugin = require("html-webpack-plugin");
const babelLoader = { // make the config reusable
  loader: "babel-loader", // name of the loader
  options: { // the specific Babel options
    presets: ["@babel/preset-env", "@babel/preset-react"],
  },
};
const tsLoader = {
  loader: "ts-loader", // name of the loader
  options: { // the specific TypeScript loader options
    transpileOnly: true,
  },
};
module.exports = {
  entry: { // defines the entry points
    app: "./src/script.tsx", // named ("app") entry point
```

```
    },
    resolve: {
      // defines what extensions to try out for resolving
      extensions: [".ts", ".tsx", ".js", ".jsx", ".json"],
    },
    module: {
      // defines the rules for transforming modules
      rules: [
        { // applied for all *.scss files
          test: /\.scss$/i,
          use: ["style-loader", "css-loader", "sass-loader"],
        },
        { // applied for all *.css files
          test: /\.css$/i,
          use: ["style-loader", "css-loader"],
        },
        { // applied for all *.js and *.jsx files
          test: /\.jsx?$/i,
          use: [babelLoader],
          exclude: /node_modules/,
        },
        { // applied for all *.ts and *.tsx files
          test: /\.tsx?$/i,
          use: [babelLoader, tsLoader],
        },
        { // applied for anything other than *.js, *.jsx, ...
          exclude: [/^$/, /\.(js|jsx|ts|tsx)$/i, /\.s?css$/i,
            /\.html$/i, /\.json$/i],
          type: "asset/resource",
        },
      ],
    },
    // defines plugins to use for extending Webpack
    plugins: [
      new HtmlWebpackPlugin({
```

```
      template: "./src/index.html",
    }),
  ],
};
```

The preceding code is quite lengthy. One of the often-heard criticisms of Webpack is that its configuration tends to become quite complex quickly.

An important part of a Webpack configuration is the use of regular expressions. The `test` and `exclude` parts of a rule work best with a regular expression. Therefore, instead of having a string with some magic behavior or a very explicit and repetitive function, a regular expression is supplied that will check if the current module should be processed by this rule or not.

The options for each loader or plugin are determined by the respective loader or plugin. Therefore, knowing Webpack is not sufficient to successfully write a Webpack configuration. You'll always need to look up the documentation of the different parts that are used in the configuration. In the preceding configuration, this has been the case with the `ts-loader` and `babel-loader` configurations.

Loaders are evaluated from right to left. For instance, in the case of `*.scss` files, the content is first transformed by `sass-loader`, which then hands over to `css-loader`. In the end, all the content is packaged up as a `style` tag by `style-loader`.

We don't always need to use a dedicated package for loaders. Using the `type` property highlighted in the previous code, we can use some premade loaders from Webpack, such as the resource loader (`asset/resource`) to return the paths to referenced files. Other options include data URIs (`asset/inline`) and accessing a file's raw content (`asset/source`).

Another way to use Webpack is to start a small server during development. Whenever we make updates to the code, the bundler can reprocess the changed parts and automatically update any active browsing sessions. All in all, this is a very convenient and quite productive way of writing a frontend application.

For Webpack's live server to work, we'll need to install yet another tooling dependency:

```
$ npm install webpack-dev-server --save-dev
```

This allows us to run the `serve` command:

```
$ npx webpack serve --mode development
<i> [webpack-dev-server] Project is running at:
<i> [webpack-dev-server] Loopback: http://localhost:8081/
<i> [webpack-dev-server] On Your Network (IPv4):
http://172.25.98.248:8081/
<i> [webpack-dev-server] Content not from webpack is served
from '/home/node/examples/Chapter06/example01/public' directory
```

```
[...]
webpack 5.74.0 compiled successfully in 1911 ms
```

The live server will keep on running until it is forcefully shut down. On the command line, this can be done by pressing *Ctrl + C*.

One thing to add to `webpack.config.js` would be the history API fallback for the development server (the `devServer` section in a Webpack configuration). This will improve the development experience of a SPA by a fair margin:

```
// ... like beforehand
module.exports = {
  // ... like beforehand
  devServer: {
    historyApiFallback: true,
  },
};
```

This setting will respond to all 404 URLs with `index.html` of the root directory – just like a SPA should be configured in production mode. This way, refreshing when being on a page with a different path than / will still work. Without the shown configuration, the 404 error will be shown in the browser – no SPA will load and handle the route.

Now that we know how bundling the example application works in Webpack, it's time to look at a more lightweight alternative named esbuild.

Using esbuild

esbuild is quite a new tool that focuses on performance. The key to esbuild's enhanced performance is that it was written from the ground up in the Go programming language. The result is a native binary that has certain advantages over pure JavaScript solutions.

If esbuild stopped at providing a native solution, it would potentially not be qualified to make this list. After all, flexibility and the option to extend the original functionality are key for any kind of bundler. Luckily, the creator of esbuild has thought about this and come up with an elegant solution. While the core of esbuild remains native – that is, written in Go and provided as a binary – plugins can be written using JavaScript. This way, we get the best of both worlds.

To get started with esbuild, we need to install the `esbuild` package using npm:

```
$ npm install esbuild --save-dev
```

With this one installation, you can use esbuild programmatically, as well as directly from the command line.

To run esbuild from the command line, you can use npx together with the esbuild executable:

```
$ npx esbuild
```

This will show all the CLI options. To do something, at least one entry point needs to be supplied:

```
$ npx esbuild --bundle src/script.tsx --outdir=dist --minify
⊠ [ERROR] No loader is configured for ".scss" files: src/
style.scss

    src/script.tsx:3:7:
      3 │ import './style.scss';
        │        ~~~~~~~~~~~~~~~


⊠ [ERROR] No loader is configured for ".mp3" files: src/sound.
mp3

    src/Audio.tsx:4:22:
      4 │ import audioPath from "./sound.mp3";
        │                       ~~~~~~~~~~~~~~

[...]
5 errors
```

As expected, we miss a few configuration steps. As with Webpack, the best way to teach esbuild about these extra bits is by creating a configuration. Unlike Webpack, we do not have a predefined configuration file – instead, the way to configure esbuild is by using it programmatically.

To do that, we must create a new file called build.js and import the esbuild package. We can use the build and buildSync functions to trigger the bundling process via esbuild.

The previous CLI command can be written programmatically like this:

build.js

```
const { build } = require("esbuild");

build({ // provide options to trigger esbuild's build
  entryPoints: ["./src/script.tsx"], // where to start from
  outdir: "./dist", // where to write the output to
```

```
  bundle: true, // bundle the resulting files
  minify: true, // turn on minification
});
```

Of course, the given script will essentially give us the same error as using the CLI directly. Therefore, let's add a few things:

- `esbuild-sass-plugin` integrates the transformation of SASS into CSS files

- `@craftamap/esbuild-plugin-html` allows us to use template HTML files

Before we can use these two plugins, we need to install them:

```
$ npm i esbuild-sass-plugin @craftamap/esbuild-plugin-html --save-dev
```

Once the plugins are installed, we can extend the `build.js` file so that it includes these two plugins:

build.js

```
const { build } = require("esbuild");
const { sassPlugin } = require("esbuild-sass-plugin");
const { htmlPlugin } = require("@craftamap/esbuild-plugin-html");

build({
  entryPoints: ["./src/script.tsx"],
  outdir: "./dist",
  format: "esm", // use modern esm format for output
  bundle: true,
  minify: true,
  metafile: true, // required for htmlPlugin
  splitting: true, // allow lazy loading
  loader: {
    ".jpg": "file",
    ".webp": "file",
    ".mp3": "file",
    ".mp4": "file",
  },
  plugins: [
```

```
        sassPlugin(),
        htmlPlugin({
          files: [
            {
              entryPoints: ["./src/script.tsx"],
              filename: "index.html",
              scriptLoading: "module", // because of esm
              htmlTemplate: "./src/index.html",
            },
          ],
        }),
      ],
    });
```

Along the way, we taught esbuild about our preference for the given file extensions. With the `loader` section, we map extensions to specific file loaders. The `file` type refers to a loader that will produce an external file. The import of that file will result in a reference to the file's relative output path.

To enable bundle splitting, the `splitting` option needs to be set. This also makes the use of the `esm` format necessary. It's the only format where esbuild knows how to produce scripts that can lazy load something. Additionally, `htmlPlugin` requires esbuild to produce a metafile to reflect the build artifacts. Therefore, the `metafile` option needs to be set to `true`.

Like Webpack, the ecosystem of esbuild is what makes this tool so flexible, yet hard to master. The options for the different plugins need to be collected from the different plugin documentation. Like the Webpack ecosystem beforehand, the quality of these plugins, as well as their maturity and the community behind them, varies a lot.

If you want to have a development server – just like the one we added to Webpack in the previous section – you can use the `serve` function, which can be imported from `esbuild`. The first argument describes server-specific settings such as the port where the service should be listening. The second argument comprises the build options – that is, the options we are supplying right now – as the only argument to the `build` function.

Let's write another script called `serve.js` to illustrate this usage:

serve.js

```
const { serve } = require("esbuild");
const { sassPlugin } = require("esbuild-sass-plugin");
const { htmlPlugin } = require("@craftamap/esbuild-plugin-
```

```
  html");

// use helper from esbuild to open a dev server
serve(
  {
    // will be reachable at http://localhost:1234
    port: 1234,
  },
  {
    // same options as beforehand (supplied to build())
    // ...
  }
);
```

One thing that esbuild does not do at the moment is HMR. Consequently, the developer's experience of just using esbuild may be a little bit underwhelming in that area when compared to similar tools.

With this in mind, let's explore yet another option that is widely used for bundling – let's have a look at the configuration-less Parcel bundler.

Using Parcel

When Parcel arrived in the community, the hype around it was massive. The reason for this was to be found in one new feature: configuration-free bundling. Parcel tried to leverage information that was already given in package.json – or configuration files written for specific tools such as Babel. Using this mechanism, the creators of Parcel thought to remove the complexity of configuring a bundler.

Ultimately, however, the whole aspect backfired in some sense. As mentioned previously, a bundler requires some flexibility. To achieve this kind of flexibility, a sound configuration system is necessary. While the configuration system of Webpack is a bit too verbose and complex, the one provided with esbuild might be a bit too low-level.

The successor of the original Parcel now also offers an optional configuration system, which tries to be right between the verbosity of Webpack and the low-level one of esbuild. This makes Parcel no longer configuration-free, but rather a configuration-less bundler.

To get started with Parcel, we need to install the parcel package using npm:

```
$ npm install parcel --save-dev
```

With this installation, you can use Parcel programmatically, as well as directly from the command line.

To run Parcel from the command line, you can use npx together with the `parcel` executable. For Parcel, the entry point can be the HTML file:

```
$ npx parcel src/index.html
```

In our case, we still need to modify the HTML so that it also points to the other source files to continue building up the module graph. A version of the `index.html` file that fits much better with Parcel would look as follows:

index.html

```
<!DOCTYPE html>
<html lang="en">
<head>
    <meta charset="UTF-8">
    <meta http-equiv="X-UA-Compatible" content="IE=edge">
    <meta name="viewport" content="width=device-width,
        initial-scale=1.0">
    <title>Bundler Example</title>
    <link rel="stylesheet" href="./style.scss">
</head>
<body>
<div id="app"></div>
<script type="module" src="./script.tsx"></script>
</body>
</html>
```

Importantly, we've added the stylesheet and script entry points. These will be detected by Parcel and properly bundled. In the end, the HTML file will be used as a template – with the entry points being replaced by the bundled stylesheet and script file references.

Starting Parcel right now will already partially work, but at this time, Parcel still has some problems with our audio and video files. While Parcel knows most image files (such as `*.webp` or `*.png`) already, some other assets need to be configured first. In Parcel, this means creating a `.parcelrc` file and adding a section about the transformers to use:

.parcelrc

```
{
    "extends": "@parcel/config-default",
```

```
  "transformers": {
    "*.{mp3,mp4}": ["@parcel/transformer-raw"]
  }
}
```

The configuration instructs Parcel to still rely on the very well-chosen defaults. However, we also added the definitions for the two file types in question to the section that handles the transformation logic. Like Webpack or esbuild, Parcel also has an in-built type to handle such imports by returning a filename that can be used within the code. In the case of Parcel, this type is called @parcel/transformer-raw.

Now, let's see if Parcel is already running:

```
$ npx parcel src/index.html
Server running at http://localhost:1234
✨ Built in 12ms
```

By default, Parcel will start a development server. This already contains everything that is needed for developing an application. Quite convenient. If we want to build the files – for example, to place the output artifacts on a server – we can use the build subcommand:

```
$ npx parcel build src/index.html
✨ Built in 6.11s

dist/index.html                402 B     4.73s
dist/index.3429125f.css      39.02 KB    149ms
dist/index.cb13c36e.js      156.34 KB    1.90s
dist/Home.bf847a6b.js         1.05 KB    148ms
dist/river.813c1909.webp     29.61 KB    150ms
dist/snow.390b5a72.jpg       13.28 KB    138ms
dist/Video.987eca2d.js        908 B      1.90s
dist/earth.4475c69d.mp4   ⚠  1.5 MB      61ms
dist/Video.61df35c5.js      611.76 KB    4.62s
dist/Audio.677f10c0.js        908 B      149ms
dist/sound.6bdd55a4.mp3     746.27 KB    92ms
```

There are CLI flags and options to set almost everything, such as the output directory. Nevertheless, by default, the quite common dist folder is chosen.

Last, but not least, let's have a look at the quite trendy Vite bundler, which tries to combine the advantages of all previous approaches into a single tool.

Using Vite

The latest addition to the set of popular bundlers is Vite. It combines a few existing tools – such as Rollup.js and esbuild – together with a unified plugin system to allow rapid development. Vite's approach is to give you the power of Webpack at the speed of esbuild.

Originally, Vite was built by the creator of the frontend framework Vue. However, as time went on, Vite's plugin system became a lot more powerful. With its increased API surface, other frontend frameworks such as React or Svelte could be supported. Now, Vite has evolved from a single-purpose tool to a real Swiss Army knife – thanks to a well-thought-out plugin mechanism with an active community.

To get started with Vite, we need to install the `vite` package using npm:

```
$ npm install vite --save-dev
```

With this installation, you can use Vite programmatically, as well as directly from the command line.

One thing to know about Vite is that it embraces having an `index.html` file as an entry point even more than Parcel. For Vite to work as intended, we need to move the `index.html` file from the `src` folder to the parent directory – that is, the project's root folder.

As we did previously, we should set the references properly:

index.html

```html
<!DOCTYPE html>
<html lang="en">
<head>
    <meta charset="UTF-8">
    <meta http-equiv="X-UA-Compatible" content="IE=edge">
    <meta name="viewport" content="width=device-width,
       initial-scale=1.0">
    <title>Bundler Example</title>
    <link rel="stylesheet" href="./src/style.scss">
</head>
<body>
<div id="app"></div>
<script type="module" src="./src/script.tsx"></script>
</body>
</html>
```

To run Vite from the command line, you can use npx together with the vite executable:

```
$ npx vite

  VITE v3.0.5  ready in 104 ms

  ➡  Local:   http://localhost:5173/
  ➡  Network: use --host to expose
```

This starts quickly as nothing has been bundled or transformed yet. Only when we hit the server will Vite start to transform things – and only the things that we are currently looking at. If you are interested in a more realistic picture, then the preview subcommand can be handy. It does a production build but exposes the outcome via the development server.

Of course, like with Parcel, we can still produce files that can be placed on a server. Very similar to Parcel, we can do this with the build subcommand:

```
$ npx vite build
vite v3.0.5 building for production...
✓ 110 modules transformed.
dist/assets/river.4a5afeaf.webp    29.61 KiB
dist/assets/snow.cbc8141d.jpg      13.96 KiB
dist/assets/sound.fa282025.mp3     746.27 KiB
dist/assets/earth.71944d74.mp4     1533.23 KiB
dist/index.html                    0.42 KiB
dist/assets/Home.82897af9.js       0.45 KiB / gzip: 0.23 KiB
dist/assets/Video.ce9d6500.js      0.36 KiB / gzip: 0.26 KiB
[...]
dist/assets/index.404f5c02.js      151.37 KiB / gzip: 49.28 KiB
dist/assets/Player.c1f283e6.js     585.26 KiB / gzip: 166.45 KiB
```

For this example, Vite is the only bundler that just works – at least once all the prerequisites have been fulfilled. If you require a custom configuration, such as for adding some plugins, then you can follow Webpack's approach and create a vite.config.js file in the project's root folder.

Now, let's recap what you've learned in this chapter.

Summary

In this chapter, you learned what a bundler is, why you need it, what bundlers exist, and how you can configure and use them. You are now able to take your web projects from their raw source code to build production-ready assets.

Equipped with detailed knowledge about bundlers, you can create very reliable code bases that are tailored toward efficiency. Not only will unnecessary code be removed upon bundling, but also all referenced files will be processed and taken into consideration. Therefore, you'll never have to worry about missing a file.

The large variety of existing bundlers can be intimidating at first. While there are some obvious choices, such as the very popular Webpack bundler, other options may be even better due to less complexity or better performance, depending on the project you have at hand. If in doubt, you can refer to the *Comparing available bundlers* section of this chapter to ascertain which bundler might be the best fit for you.

In the next chapter, we will take closer look at another category of crucial development tools. We'll see how testing tools give us confidence that our code works as it should, both today and in the future.

7

Improving Reliability with Testing Tools

Now that we can actually write and build our code for the browser efficiently, it makes sense to also consider verifying the code's output. Does it really fulfill the given requirements? Has anything changed in terms of the expected outcome? Does the code crash when unexpected values are passed in?

What we need to answer these questions is testing. Testing can mean a lot of things – and depending on who you ask, you'll get a different answer to the question "What should we test?" In this chapter, we'll walk through the different options that interest us as developers. We'll see what tools exist to automate these tests and how we can set them up and use them practically.

We will start our journey into the testing space with a discussion on the beloved testing pyramid. We will then continue by learning about the types of test tools – most notably, pure runners and whole frameworks. Finally, we'll cover some of the most popular tools in this space.

By the end of this chapter, you will know which testing framework or test runner to choose for your programming needs, along with the pros and cons of each option.

We will cover the following key topics in this chapter:

- Considering the testing pyramid
- Comparing test runners versus frameworks
- Using the Jest framework
- Using the Mocha framework
- Using the AVA test runner
- Using Playwright for visual tests
- Using Cypress for end-to-end testing

Technical requirements

The complete source code for this chapter is available at `https://github.com/PacktPublishing/Modern-Frontend-Development-with-Node.js/tree/main/Chapter07`.

The CiA videos for this chapter can be accessed at `https://bit.ly/3DW9yoV`.

Considering the testing pyramid

Over the years, more and more types of software testing have been identified and added to the standard repertoire of software projects and testing professionals such as quality assurance engineers. A powerful tool to categorize and order the most common types of software testing is the testing pyramid.

The testing pyramid arranges the different types of testing by their visibility and effort. Higher layers of the pyramid require more effort but have greater visibility. Tests that are placed in the lower layers of the pyramid should be written a lot more – after all, these are the foundations of the pyramid.

An illustration of the testing pyramid is shown in *Figure 7.1*. The basis of the testing pyramid is formed by unit tests, which provide enough reliability to run components and integration tests on top of them later. Finally, UI tests (quite often referred to as end-to-end tests) can be run to verify that the solution works for end users:

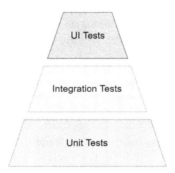

Figure 7.1 – The testing pyramid with three layers of automatic testing

Usually, end-to-end tests refer to tests that use the interface as presented to the end user. In the case of a web application, this would be the actual website. By their nature, end-to-end tests are usually black-box tests. The whole system is treated as is, so with everything running as close to the production environment as possible.

> **Black-box tests**
>
> The notion of a black box comes from the so-called black box approach. This is a common technique to analyze an open system by varying the input and measuring the output. This approach makes sense when the internal workings are either not known or not accessible. Likewise, black-box testing is also performed without changing the application's internal workings.

Variations of end-to-end tests focus on performance (*load tests*) or security (*penetration tests*). While the former can be quite tricky and expensive to run, the latter should be performed regularly to shield against potential attacks. One of the greatest risks for companies is to be hacked. Not only will this include the theft of precious data but it will also have a strong negative impact on the company's brand. To help defend against scenarios like this, sometimes gray-box testing is used, which, unlike black-box tests, understands certain documented operations of the system.

One challenge with testing is that many of the terms used, such as integration or component tests, are not universally defined. For instance, some people consider an integration test to be something very narrow – testing the integration with one external part at a time. Other people may say that an integration test should cover the integration with all the external parts. Consequently, it is quite important to carefully review and define these terms before using them in a project.

When we refer to unit tests, we mean tests for a single unit (such as a function) of the project – only one specific part of it. Usually, this unit carries some logic that can be tested specifically. Everything that is not related to this one unit has to be controlled. While some unit tests can be written like black-box tests, most unit tests will require a detailed understanding of the internal workings. This way, the behavior of the tested unit can be controlled as required.

Consider the following code:

---

pure.js

```
export function pickSmallestNumber(...numbers) {
  if (numbers.length > 0) {
    return numbers.reduce(
        (currentMin, value) => Math.min(currentMin, value),
        Number.MAX_VALUE);
  }

  return undefined;
}
```

In the preceding code, the function is very well suited for a unit test:

- It is exported, so we can access it from another module containing the tests.
- It does not use anything outside of the function – it's a so-called pure function.
- The logic is sufficiently complex to test against a set of predefined test cases.

Unit tests for the `pickSmallestNumber` function could look as follows:

```
test('check if undefined is returned for no input', () => {
  const result = pickSmallestNumber();
  assert(result === undefined);
});

test('check if a single value is the smallest number',
  () => {
  const result = pickSmallestNumber(20);
  assert(result === 20);
});

test('check if 1 is smaller than 5', () => {
  const result = pickSmallestNumber(5, 1);
  assert(result === 1);
});

test('check if -1 is smaller than 0 but larger than -5',
  () => {
  const result = pickSmallestNumber(-1, -5, 0);
  assert(result === -5);
});
```

> **Note**
>
> As mentioned, the code could look like this. The functions used are defined nowhere and the preceding code would not run as presented.

For these tests, we introduced a new function, `test`, which takes a description of the test and the code in the form of a function for running the test. We also introduced an assertion function, `assert`, which could be taken from the Node.js built into the `assert` module. The proposed `assert` function accepts a Boolean input – throwing an exception if the input is `false`. The testing tools and frameworks we'll look at will replace these constructs with more expressive and elegant alternatives.

Besides the actual testing and test area differences, the tooling choices also offer a few variations. One of the most crucial ones is the difference between a full testing framework and a test runner.

Comparing test runners versus frameworks

Historically, tests for JavaScript targeting web browsers could not be just written and run automatically. The main reason was that this involved dealing with a real browser. There was no way to just *pretend to run in the browser*. For this reason alone, the first tools in that space have either been scripts or whole websites evaluating JavaScript or browser automation tools. The latter actually forms its own category – being at the heart of modern end-to-end tests.

The main driver for running the tests – historically, for starting everything that needs to be running to actually perform tests – is called a test runner. One of the first very successful test runners in the JavaScript space was **Karma**. The job of Karma was to spin up a server that runs a website hosting the tests, which are targeting JavaScript code that should run in a browser. Karma then opened available browsers to access the hosted website running the tests. The results were reported back to the server and shown in the console.

If all this sounds complicated to you – you would be right, it was. The job of these runners was to make this process as reliable as possible. They also tried to be user-friendly and hide the underlying complexity as much as possible.

Today, test runners like Karma are not really necessary. Instead, most test runners such as **AVA** stay in the console by leveraging Node.js. When JavaScript code requires the browser API, which is most like the DOM API, the runner just emulates these missing APIs. As a result of the emulation, the JavaScript code that is tested can run as it would in the browser, but everything remains in Node.js.

While the part about emulating the DOM API sounds great, it is actually not within the scope of a test runner. Test runners are really only focused on running the tests. Instead, developers establish the emulation part somewhat or pick a full test framework. A full test framework should already have figured out things such as the DOM API emulation so that they can be easily added, or they are already part of the standard installation.

A full test framework not only includes a test runner but also things such as an assertion library. So far, we've only used some kind of `assert` function with a proposed behavior. A full assertion library would give us a set of functions that makes the debugging process in the case of a failed assertion quite easy. Already from the test output, we would see which assertion was broken – and why.

An example of a good assertion library is **Chai**. It comes with three different exports: `should`, `expect`, and `assert`. The export that you see most often used in test code is `expect`.

Using `expect` from the `chai` package, the first two test cases from our preceding unit tests could be rewritten as follows:

```
test('check if undefined is returned for no input', () => {
  const result = pickSmallestNumber();
  expect(result).to.be.undefined;
});
```

```
test('check if a single value is the smallest number',
  () => {
  const result = pickSmallestNumber(20);
  expect(result).to.equal(20);
});
```

The beauty of the rewritten code is that it almost reads like text. Even somebody with less experience in the testing framework, JavaScript, or Node.js could identify what the test does – and even more importantly – what it tries to verify. The chaining of the expectation using the member (.) operator is one of the things that makes Chai such a popular assertion library.

Every testing framework comes with an assertion library. Some frameworks may even let the user decide which assertion library to use.

Now that we know all the basics of testing JavaScript-based applications, we should explore some of the tools to actually implement such tests. We will start with one of the most commonly used testing utilities: the Jest test framework.

Using the Jest framework

Jest is a modern test framework that was authored by Facebook to fully leverage Node.js for running tests. It should have the power to run all the tests required at Facebook without requiring a diploma in engineering to understand, control, or modify it.

To use Jest, you need to install the jest package from npm:

```
$ npm install jest --save-dev
```

This allows you to use the jest command-line utility. Ideally, run it with npx as we did with the other tools:

```
$ npx jest
```

Jest can be configured by providing a jest.config.js file. The easiest way to create this kind of file is by using the jest tool with the --init flag. This will guide us through some questions to create a suitable configuration:

```
$ npx jest --init

The following questions will help Jest to create a suitable
configuration for your project
```

☑ Would you like to use Jest when running "test" script in "package.json"? … yes

☑ Would you like to use Typescript for the configuration file? … no

☑ Choose the test environment that will be used for testing > jsdom (browser-like)

☑ Do you want Jest to add coverage reports? … no

☑ Which provider should be used to instrument code for coverage? > v8

☑ Automatically clear mock calls, instances, contexts and results before every test? … yes

▨ Modified /home/node/example/Chapter07/package.json

▤ Configuration file created at /home/node/example/Chapter07/jest.config.js

In this case, we've instructed Jest to change the `test` script in `package.json`. Now, when we run `npm run test` or just `npm test` in our terminal for the current project, Jest will start. The options for the test environment and coverage are interesting to us.

Let's have a look at the essential parts of the generated configuration file:

```
module.exports = {
    clearMocks: true,
    coverageProvider: "v8",
    testEnvironment: "jsdom",
};
```

The generated configuration file also contains a lot of comments and commented-out options. This way, you can configure Jest without having to consult the official documentation website.

The given configuration has just one problem… The selected `jsdom` environment only works when a special package called `jest-environment-jsdom` is installed. This has been changed in version 28 of Jest and is, unfortunately, not done automatically:

```
$ npm install jest-environment-jsdom --save-dev
```

Luckily, the error messages in Jest are usually quite good and very helpful. Even without knowing these things, we'll get proper messages that tell us exactly what to do.

One last thing we should consider is using Babel for code transformations. These transformations are unnecessary if we write pure Node.js-compatible code (such as by using CommonJS). Otherwise, code transformations are necessary. In general, Jest uses code transformations to make any kind of used code – not only plain JavaScript but also flavors such as TypeScript and Flow – usable without requiring special treatment upfront.

First, let's install the `babel-jest` plugin and the required `@babel/core` package:

```
$ npm install babel-jest @babel/core @babel/preset-env --save-
dev
```

Now, let's extend `jest.config.js` with the `transform` configuration section:

```
module.exports = {
  // as beforehand
  "transform": {
    "\\.js$": "babel-jest",
  },
};
```

The new section tells Jest to use the `babel-jest` transformer for all files ending with `.js`. Also add a `.babelrc` file as discussed in *Chapter 4, Using Different Flavors of JavaScript*:

```
{
  "presets": ["@babel/preset-env"]
}
```

With this configuration, Babel will properly transform the given files. The test code can now be written as follows:

pure.test.js

```
import { pickSmallestNumber } from "./pure";

it("check if undefined is returned for no input", () => {
  const result = pickSmallestNumber();
  expect(result).toBeUndefined();
});


it("check if a single value is the smallest number", () => {
```

```
  const result = pickSmallestNumber(20);
  expect(result).toBe(20);
});

it("check if 1 is smaller than 5", () => {
  const result = pickSmallestNumber(5, 1);
  expect(result).toBe(1);
});

it("check if -1 is smaller than 0 but larger than -5",
  () => {
  const result = pickSmallestNumber(-1, -5, 0);
  expect(result).toBe(-1);
});
```

While Jest also supports a `test` function as in our pseudo implementation introduced in the *Considering the testing pyramid section*, the `it` function is much more commonly seen. Note that Jest comes with its own integrated assertion library, which uses the `expect` function. The `expect` function is also called a **matcher**.

> **Matchers**
>
> For our simple example, the matcher will only have to deal with strings and numbers. In general, however, any kind of JavaScript input, such as arrays or objects, can be matched and asserted. The `expect` function has some helpers to deal with, for instance, object equality (`toBe`), as in, having the same reference, and equivalence (`toEqual`), as in, having the same content.

Let's run this:

```
$ npm run test

> Chapter07@1.0.0 test /home/node/example/Chapter07
> jest

  PASS  src/pure.test.js
    ✓ check if undefined is returned for no input (2 ms)
    ✓ check if a single value is the smallest number (1 ms)
    ✓ check if 1 is smaller than 5
    ✓ check if -1 is smaller than 0 but larger than -5
```

```
Test Suites: 1 passed, 1 total
Tests:       4 passed, 4 total
Snapshots:   0 total
Time:        0.818 s, estimated 1 s
Ran all test suites.
```

Great – our code works. By default, Jest will look for all files ending with `.test.js`. By convention, `.spec.js` files would also work. The convention used can be changed though.

Today, Jest is arguably the most used testing framework. However, especially older projects potentially use something else. A very solid and common occurrence here is Mocha. Like Jest, it is also a testing framework, but with a few key differences.

Using the Mocha framework

Mocha is an older but feature-rich testing framework that runs in Node.js and also the browser. In this section, we'll exclusively use Mocha in Node.js. Unlike Jest, the notion of an environment does not exist. Nevertheless, a similar setup can be achieved, where browser APIs would be emulated by some npm package such as `jsdom`.

To use Mocha, you need to install the `mocha` package from npm:

```
$ npm install mocha --save-dev
```

This allows you to use the `mocha` command-line utility. Ideally, run it with `npx` as we did with the other tools:

```
$ npx mocha
```

At this point, not much is working. By default, Mocha follows a different convention from Jest. Here, we need to specify a different pattern or place our tests in a folder named `test`.

What we definitely need to do is to include Babel for code transformations. This works a bit differently than with Jest. Instead of a dedicated plugin, we only integrate the `@babel/register` package, which will automatically transform any code when a module is loaded:

```
$ npm install --save-dev @babel/register @babel/core @babel/
preset-env
```

Now, we can copy the `.babelrc` file that we used previously with Jest. For Mocha, the configuration can be placed in a file called `.mocharc.js`. Setting up the configuration file to always require the `@babel/register` package first looks like this:

.mocharc.js

```js
module.exports = {
  require: "@babel/register",
};
```

Mocha is a kind of special testing framework, as it does not come with an assertion library. Instead, it relies on other assertion libraries. As long as it throws an exception in case of a mismatch, the assertion works.

To write tests with Mocha without using a special assertion library besides the one that already comes with Node.js, we would write our tests as follows:

pure.test.js

```js
import { equal } from "assert";
import { pickSmallestNumber } from "../src/pure";

it("check if undefined is returned for no input", () => {
  const result = pickSmallestNumber();
  equal(result, undefined);
});

it("check if a single value is the smallest number", () => {
  const result = pickSmallestNumber(20);
  equal(result, 20);
});

it("check if 1 is smaller than 5", () => {
  const result = pickSmallestNumber(5, 1);
  equal(result, 1);
});

it("check if -1 is smaller than 0 but larger than -5",
  () => {
  const result = pickSmallestNumber(-1, -5, 0);
  equal(result, -5);
});
```

In the preceding code, the `it` functions follow the same behavior as in Jest.

Now, let us run `mocha` via `npm test`:

```
$ npm run test


> example02@1.0.0 test /home/node/example/Chapter07/example02
> mocha


   ✓ check if undefined is returned for no input
   ✓ check if a single value is the smallest number
   ✓ check if 1 is smaller than 5
   ✓ check if -1 is smaller than 0 but larger than -5


   4 passing (3ms)
```

Compared to Jest, we get a little bit less output. Still, all the relevant information is presented and if there were an error, we would have gotten all the necessary information to identify and fix the issue. The crucial difference between Jest and Mocha is that Jest really breaks down the tests according to their associated test module, while Mocha just presents the results.

Mocha is actually quite feature-packed and everything but lightweight. A more streamlined option is to avoid using a full testing framework and instead go for a test runner only. One option is to use AVA.

Using the AVA test runner

AVA is a modern test runner for Node.js. It stands out because of its ability to embrace new JavaScript language features and cutting-edge properties of Node.js, such as process isolation. In this way, AVA executes tests very quickly and reliably.

To use AVA, you need to install the `ava` package from npm:

```
$ npm install ava --save-dev
```

This allows you to use the `ava` command-line utility. Ideally, run it with `npx` as we did with the other tools:

```
$ npx ava
```

While Mocha and Jest could also be installed globally, AVA only works in projects as a local dependency. As this is the better setup anyway, there should be no practical downside from this constraint.

As mentioned, AVA is built quite closely on Node.js – following its conventions and rules wherever possible. In this regard, AVA also allows us quite quickly to adapt ESM instead of CommonJS. By modifying package.json for the project, we get immediate support for using ESM in our tests, too:

package.json

```
{
  // like beforehand
  "type": "module",
  // ...
}
```

By default, AVA looks for files that follow the same pattern as Jest. Therefore, files that end with .test.js and .spec.js will be found among others. There is no need to configure AVA or place the tests in a separate directory.

The other thing that AVA does is to provide a function as a default export from the ava package. This function is needed to declare tests. Each test then receives a so-called test context as a callback parameter for its implementation. This way, AVA feels a lot more explicit and less magical than the other solutions.

Let's see how we can write the tests with AVA:

pure.test.js

```
import test from 'ava';
import { pickSmallestNumber } from "./pure.js";

test("check if undefined is returned for no input", (t) => {
  const result = pickSmallestNumber();
  t.is(result, undefined);
});

test("check if a single value is the smallest number",
  (t) => {
  const result = pickSmallestNumber(20);
  t.is(result, 20);
});
```

```
test("check if 1 is smaller than 5", (t) => {
  const result = pickSmallestNumber(5, 1);
  t.is(result, 1);
});

test("check if -1 is smaller than 0 but larger than -5",
  (t) => {
  const result = pickSmallestNumber(-1, -5, 0);
  t.is(result, -5);
});
```

Overall, the structure is similar to the previous two full frameworks. Still, AVA is just a runner and misses things such as special assertion libraries, options for mocking, and snapshots, among other things.

To run the tests, we can adjust the `test` script in `package.json`. Triggering the `ava` utility, a run with the AVA test runner looks like this:

```
$ npm run test

> example03@1.0.0 test /Users/node/example/Chapter07/example03
> ava


  ☑ check if undefined is returned for no input
  ☑ check if a single value is the smallest number
  ☑ check if 1 is smaller than 5
  ☑ check if -1 is smaller than 0 but larger than -5
  ─


  4 tests passed
```

Now that we covered three tools to run some code-centric tests, let's explore some options for running UI tests, too. We will start with **Playwright**, which is a modern library to automate the behavior of web browsers such as Google Chrome or Firefox.

Using Playwright for visual tests

Node.js is not only a great basis for running logical tests but also for verifying visuals, such as those of a website running in a browser. A modern approach for browser automation is Playwright.

To use Playwright, you need to install the `playwright` package from npm:

```
$ npm install playwright --save-dev
```

The `playwright` package enables you to use Playwright in an existing application, which could also be inside existing tests such as unit tests executed with Jest using the `jest-playwright-preset` package.

An even better setup can be achieved by using the `@playwright/test` test runner package:

```
$ npm install @playwright/test --save-dev
```

This allows you to use the `playwright` command-line utility. Ideally, run it with `npx` as we did with the other tools:

```
$ npx playwright test
```

Running this will look for all files matching the same conventions as previously noted in the Jest and AVA sections. Every file ending with `.test.js` or `.spec.js` will be included. Additionally, the Playwright test runner is also capable of evaluating TypeScript files. The runner therefore also includes `.test.ts` and `.spec.ts` files in its default lookup.

Let's look at a simple test run again. We'll run tests against a public website available at `https://microfrontends.art`. The test would work against a local website running on localhost, too:

mf.test.ts

```
import { test, expect } from '@playwright/test';

test('homepage has micro frontends in the title and in an
  h1', async ({ page }) => {
  await page.goto('https://microfrontends.art/');

  // Expect the title "to contain" a substring.
  await expect(page).toHaveTitle(/Micro Frontends/);

  // Grab an element ("h1")
  const h1 = page.locator('h1');

  // Expect the element to have a specific text
  await expect(h1)
    .toHaveText('The Art of Micro Frontends');
});
```

The structure feels a bit similar to AVA. As with AVA, we are using explicit imports to create the test infrastructure. We also need to use the parameter of the test's callback to actually do something useful with the website using the `page` object.

Let's change the `test` script in `package.json` and run the test provided:

```
$ npm run test

> example04@1.0.0 test /Users/node/example/Chapter07/example04
> playwright test



Running 1 test using 1 worker

  ✅  1 tests/mf.test.ts:3:1 > homepage has Playwright in title
and get started link linking to the intro page (491ms)



    1 passed (5s)
```

Yet another option to write end-to-end tests is Cypress. This promises to be even more convenient and also equipped to test individual components, too.

Using Cypress for end-to-end testing

Cypress is a focused, end-to-end testing framework that also comes with the ability to test individual UI components. It tries to be different by mostly avoiding browser automation. Instead, its test runner is located directly inside the browser.

To use Cypress, you need to install the `cypress` package from npm:

```
$ npm install cypress --save-dev
```

This allows you to use the `cypress` command-line utility. Ideally, run it with `npx` as we did with the other tools:

```
$ npx cypress open
```

Cypress is at its heart a graphical tool. As such, we are first introduced to a small configurator that allows us to set up our project. The configurator is shown in *Figure 7.2*. Picking **E2E Testing** will give you the ability to influence what files are written:

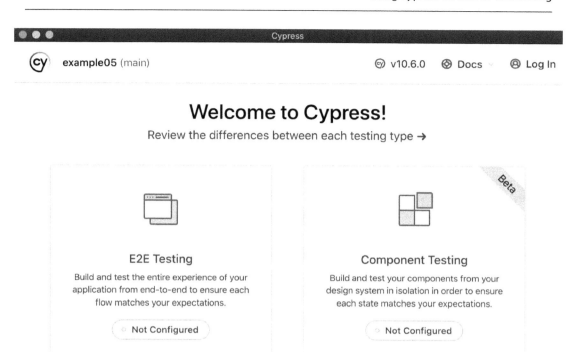

Figure 7.2 – The Cypress configurator on opening it for the first time

The configurator also lets you pick a browser where the tests should actually be run. Right now, **Chrome**, **Edge**, **Electron**, and **Firefox** are supported.

At this time, we can add our first test – in the context of Cypress, referred to as a spec or specification. We'll use the same kind of test that we've added as an example for Playwright:

mf.cy.js

```
describe("empty spec", () => {
  it("passes", () => {
    cy.visit("https://microfrontends.art");

    // Expect the title "to contain" a substring.
```

```
        cy.title().should("contain", "Micro Frontends");

        // Expect the h1 element to have a specific text.
        cy.get("h1").should("have.text",
          "The Art of Micro Frontends")
    });
});
```

As seen in the preceding small test, the whole test structure is implicit. The main downside of this is that there is no good IDE support to help with proper typing – that is, type information that can be used by TypeScript. A good way out of it is to install the `typescript` package in the project and create a `tsconfig.json` that teaches TypeScript about Cypress:

tsconfig.json

```
{
  "compilerOptions": {
    "target": "es5",
    "lib": ["es5", "dom"],
    "types": ["cypress", "node"]
  },
  "include": ["**/*.ts"]
}
```

Now, you can rename the test file to end with `.ts` (in our example, `mf.cy.ts`) and enjoy improved autocompletion in most editors and IDEs.

Running this test will yield a graphical result. In *Figure 7.3*, you can see the output from running the test in the selected browser. This is the key point of Cypress. An end-to-end test never leaves the visual area and allows us to directly interact with the test within its visual boundaries. This makes tests written with Cypress not only very beginner-friendly but also quite easy to debug:

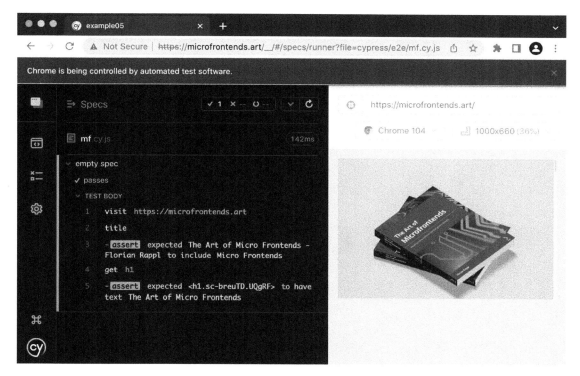

Figure 7.3 – Running the test directly in the browser

If you want to run the locally available tests directly and without visual interaction, then you can also use the run command:

```
$ npx cypress run
```

This is especially handy in non-local environments, such as a CI/CD pipeline for validating software builds.

With this in mind, let's recap what we learned in this chapter.

Summary

In this chapter, you learned about which different types of testing we can automate and how important these types are for software projects to succeed. You've seen the popular tools that exist to help us cover our projects. By following the testing pyramid, you should be able to decide what tests you need to focus on to make your project as reliable as possible.

By using the power test frameworks such as Jest or Mocha or a flexible runner such as AVA, you can automate a lot of different things – from unit tests to full end-to-end tests. Dedicated end-to-end test frameworks such as Playwright or Cypress also come with their own runners – which makes sense for complex visual tests in particular. In the unit and integration testing space, Jest comes in handy. It also allows us to quickly integrate other flavors of JavaScript or customize a lot of different features.

In the next chapter, we will finally also publish our own packages – to the public registry and other custom registries.

Part 3:
Advanced Topics

In this part, you'll dive into advanced topics such as publishing your own npm packages and structuring your projects into a shared code base such as a monorepo. You'll see what options exist and how tools such as Nx, Lerna, or Turbo can help you set up projects that can scale.

To round off your knowledge about Node.js and its ecosystem, this part will also teach you how to make use of any kind of code compiled as WebAssembly within Node.js, as well as which other runtimes can be used as the basis for web development tooling.

This part of the book comprises the following chapters:

- *Chapter 8, Publishing npm Packages*
- *Chapter 9, Structuring Code in Monorepos*
- *Chapter 10, Integrating Native Code with WebAssembly*
- *Chapter 11, Using Alternative Runtimes*

8

Publishing npm Packages

Before now, our main focus has been to learn everything about improving and contributing to existing projects, but quite often, this is not everything. Some projects will need to be initiated correctly by you and one part of this process is to decide which packages should actually be reused.

We've already learned that reusability in Node.js is primarily gained through the module system, which can be enhanced by third-party dependencies in the form of npm packages. In this chapter, you'll learn how you can publish npm packages yourself. This way, a functionality implemented once can be shared among the team working on the same project or with anyone.

To achieve our goal in this chapter, first, we'll set up a simple library to serve our case well. Then, we publish this library to the official npm registry in a way that makes the code available to any Node.js developer. If you want to keep your library a bit less exposed, then the following sections will be interesting for you. In these, you will first learn how to select other registries before you actually select a local registry to use for publishing and installation.

Finally, we'll also look at ways to broaden the scope of our library – by making it **isomorphic** or exposing it as a tool. In summary, we'll cover the following key topics in this chapter:

- Publishing to the official registry
- Selecting another npm registry via `.npmrc`
- Setting up Verdaccio
- Writing isomorphic libraries
- Publishing a cross-platform tool

Technical requirements

The complete source code for this chapter is available at `https://github.com/PacktPublishing/Modern-Frontend-Development-with-Node.js/tree/main/Chapter08`.

The CiA videos for this chapter can be accessed at `https://bit.ly/3UmhN4B`.

Publishing to the official registry

Let's start by creating a small library that uses a structure that can be seen very often in Node.js projects. The structure consists of an `src` folder, where the original sources are located, and a `lib` folder, containing the output to be used by the target system. The target system could either be something such as a bundler for browser applications or a specific version of Node.js.

To initialize this kind of project, we can use the `npm` command-line utility as we did before:

```
$ npm init -y
```

Now, we'll set everything up. First, we will install `esbuild` as a development dependency. This can be very helpful for transforming our source files into usable library files:

```
$ npm install esbuild --save-dev
```

Next, we change `package.json` to fit our needs:

---

package.json

```
{
  "name": "lib-test-florian-rappl",
  "version": "1.0.0",
  "description": "Just a test library",
  "keywords": [],
  "author": "Florian Rappl",
  "license": "MIT",
  "main": "lib/index.js",
  "source": "src/index.js",
  "scripts": {
    "build": "esbuild src/*.js --platform=node --outdir=lib
      --format=cjs"
  },
  "devDependencies": {
    "esbuild": "^0.15.0"
  }
}
```

Importantly, replace the chosen placeholder's name (`florian-rappl` in the name field and `Florian Rappl` in the `author` field) with your name. For the name field, make sure to only use letters allowed for package name identifiers. Also, feel free to change the selected license.

> **Licenses**
>
> An important piece of information in every package.json is the license field. While the MIT License is a very good choice for many open-source projects, it is by no means the only one. Other popular choices include the Apache License 2.0, BSD 3-Clause, and the ISC License.

Now, we'll add some content to our source file:

src/index.js

```
import { readFile } from "fs";
import { resolve } from "path";

export function getLibName() {
  const packagePath = resolve(__dirname,
    "../package.json");
  return new Promise((resolve, reject) => {
    readFile(packagePath, "utf8", (err, content) => {
      if (err) {
        reject(err);
      } else {
        const { name, version } = JSON.parse(content);
        resolve(`${name}@${version}`);
      }
    });
  });
}
```

This file was written in a way that makes sense for us as developers, but cannot be run by Node.js directly. The problem is twofold. First, we are using ESM syntax without guaranteeing that Node.js supports this. Second, we are mixing ESM constructs such as import and export with CommonJS constructs such as __dirname.

Luckily, we already installed esbuild to take care of this, with the defined build script actually using it for convenience:

```
$ npm run build

> lib-test-florian-rappl@1.0.0 build /home/node/code/example01
> esbuild src/*.js --platform=node --outdir=lib --format=cjs
```

```
lib/index.js  1.4kb
```

⚡ `Done in 2ms`

At this point, we have two directories in our project: `src`, containing the original sources, and `lib`, containing the CommonJS output. This is also reflected in `package.json`, where the source field points to `src/index.js` and the `main` field points to `lib/index.js`.

Just as a reminder: the `main` field tells Node.js what module to use in case the package is included via `require` – for example, `require('lib-test-florian-rappl')` would reference and evaluate the `lib/index.js` file.

Let's say you want to publish this package now to the official npm registry. For this, you first need an account on `npmjs.com/signup`. Once successfully registered and logged in, you should see a view similar to that in *Figure 8.1*:

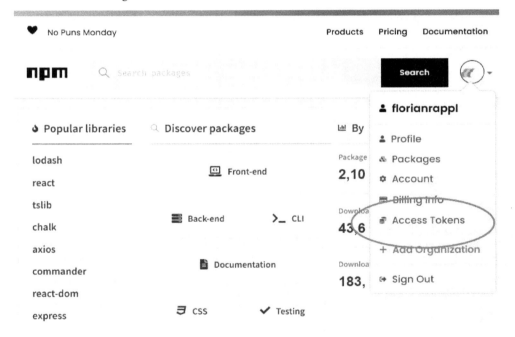

Figure 8.1 – The view on npmjs.com once logged in

On your own machine, you can now authenticate to the official npm registry by running the following:

```
$ npm login
```

This will request your username and password. Alternatively, you could authenticate using so-called access tokens. This is especially useful for scripts, such as automation running in a CI/CD pipeline. To generate a new access token, follow the link highlighted in *Figure 8.1*.

Now that you have authenticated the npm utility, you can go ahead and publish your package:

```
$ npm publish
npm notice
npm notice        lib-test-florian-rappl@1.0.0
npm notice === Tarball Contents ===
npm notice 1.5kB lib/index.js
npm notice 425B   src/index.js
npm notice 344B   package.json
npm notice === Tarball Details ===
npm notice name:          lib-test-florian-rappl
npm notice version:       1.0.0
npm notice package size:  1.1 kB
npm notice unpacked size: 2.3 kB
npm notice
shasum:          2b5d224949f9112eeaee435a876a8ea15ed3e7cd
npm notice integrity:     sha512-cBqlczwmN4vep[...]/
vXrORFGjRjnA==
npm notice total files:   3
npm notice
+ lib-test-florian-rappl@1.0.0
```

This will package your project as a compressed archive. Then, the utility will upload the tarball to the official npm registry.

Now, you can go to npmjs.com to look for your package name. You should see the package info page similar to *Figure 8.2* with more details about the published package. Note that we did not include a README.md or any keywords:

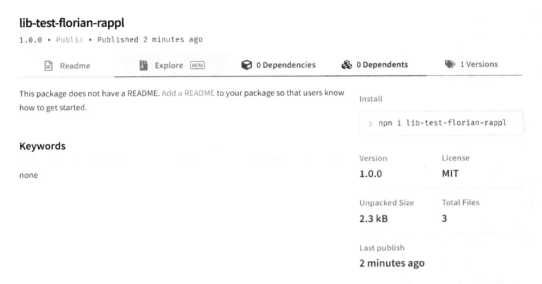

lib-test-florian-rappl

1.0.0 • Public • Published 2 minutes ago

📄 Readme	📘 Explore [BETA]	📦 0 Dependencies	🐙 0 Dependents	🏷️ 1 Versions

This package does not have a README. Add a README to your package so that users know how to get started.

Keywords

Install

> npm i lib-test-florian-rappl

Version	License
1.0.0	MIT

Unpacked Size	Total Files
2.3 kB	3

Last publish

2 minutes ago

Figure 8.2 – The details of the published package

One thing that you might consider is to give your package a scope. When you publish a package with a scope, then you'll need to configure the access settings of the package. By default, non-scoped packages are public, and scoped packages are private.

For publishing a scoped package to the official npm registry, you'll first need to be either a member or owner of an organization on the npm website. The organization name must match the name of the scope.

> **Package scope**
>
> A good way to group packages is to put them in a common scope. The scope has to start with an "@" symbol, which is followed by the name of the scope. The rules for the name of the scope are identical to package names. Besides grouping packages, scopes can be used to place certain packages in a different registry without much trouble. Most importantly, scopes can be reserved on the official npm registry, such that only authorized accounts can publish new packages using a reserved scope.

To consistently publish a scoped package such as @foo/bar with public access, you need to modify the package.json. The relevant configuration is stored in a property called publishConfig:

package.json

```
{
    "name": "@foo/bar",
    // ... like beforehand
```

```
  "publishConfig": {
    "access": "public"
  }
}
```

Alternatively, the access configuration could also be set directly when using the npm publish command with the --access=publish flag.

So far, we have only discussed how we can publish something to the official npm registry. What about choosing some other npm registry? For this, we need to change the .npmrc file.

Selecting another npm registry via .npmrc

To configure the behavior of npm, a special file called .npmrc is used. We've already briefly touched on this file in *Chapter 3, Choosing a Package Manager*. This file can be used not only to determine the source of the packages but also to define where to publish to.

A simple modification might look as follows:

.npmrc

```
; Lines starting with a semicolon or
# with a hash symbol are comments
registry=https://mycustomregistry.example.org
```

This way, all installations and publish attempts will be performed at https://mycustomregistry.example.org instead of the official registry located at https://registry.npmjs.org.

Quite often, this extreme approach is unnecessary or even unwanted. Instead, you might only want to use another registry for a subset of the packages. In the most common case, the subset is already defined by a scope.

Let's say the @foo scope that we used in the previous section with the @foo/bar package should be bound to a custom registry, while all the other packages can still be resolved by the official one. The following .npmrc covers this:

.npmrc

```
@foo:registry=https://mycustomregistry.example.org
```

While the local `.npmrc` – that is, the one adjacent to a `package.json` of a project – should be used to define the registries, a global `.npmrc` – located in your home directory – should be used to provide information regarding authentication. Quite often, a private registry can only be used with such authentication information:

~/.npmrc

```
//mycustomregistry.example.org/:username="myname"
//mycustomregistry.example.org/:_password="mysecret"
//mycustomregistry.example.org/:email=foo@bar.com
always-auth=true
```

The `always-auth` setting is used to tell npm that even `GET` requests – that is, requests for resolving or downloading packages – need to use the provided authentication.

An easy way to test custom configuration is to roll out your own npm registry. A good way of doing that locally is to use the open source project **Verdaccio**.

Setting up Verdaccio

There are a couple of commercial registry options out there. Arguably, the most popular option is to get a pro plan for the official npm registry. This way, you'll be able to publish and manage private packages. Whatever option you pick, you will always have to use a cloud version for publishing your packages.

Especially for playing around with the publishing process, having a registry locally would be great. A great option is to leverage **Verdaccio** for this. Verdaccio can be either run by cloning the Verdaccio code repository, running the **Docker** container provided by Verdaccio, or using npx.

Let's go for the npx approach:

```
$ npx verdaccio
 warn --- config file  - ~/.config/verdaccio/config.yaml
 info --- plugin successfully loaded: verdaccio-htpasswd
 info --- plugin successfully loaded: verdaccio-audit
 warn --- http address - http://localhost:4873/ -
verdaccio/5.14.0
```

Now that Verdaccio is running, you can go to the URL shown in the console. You should see Verdaccio's home page as shown in *Figure 8.3*:

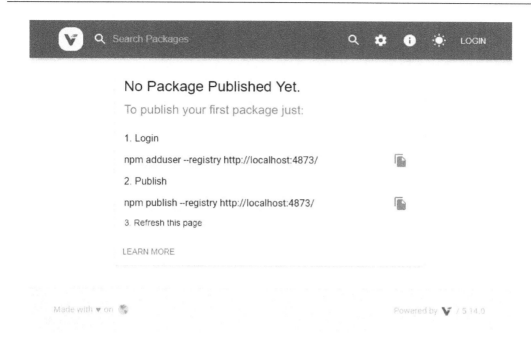

Figure 8.3 – The home page of Verdaccio with publishing instructions

Let's say we want to publish the package we created earlier to Verdaccio instead of the official npm registry. The steps we need to follow are these:

1. Authenticate against the new registry (in Verdaccio, you can use whatever credentials you'd like by default, but npm requires you to authenticate)

2. Either configure the URL to your running instance of Verdaccio via a .npmrc file or by explicitly using the --registry flag with the npm publish command

In practice, these two steps look as follows:

```
$ npm adduser --registry http://localhost:4873/
Username: foo
Password:
Email: (this IS public) foo@bar.com
Logged in as foo on http://localhost:4873/.
$ npm publish --registry http://localhost:4873
npm notice
npm notice     lib-test-florian-rappl@1.0.0
npm notice === Tarball Contents ===
```

```
npm notice 1.5kB lib/index.js
npm notice 425B  src/index.js
npm notice 344B  package.json
npm notice === Tarball Details ===
npm notice name:          lib-test-florian-rappl
npm notice version:       1.0.0
npm notice package size:  1.1 kB
npm notice unpacked size: 2.3 kB
npm notice
shasum:       2b5d224949f9112eeaee435a876a8ea15ed3e7cd
npm notice integrity:     sha512-cBq1czwmN4vep[...]/
vXrORFGjRjnA==
npm notice total files:   3
npm notice
+ lib-test-florian-rappl@1.0.0
```

Once published, the package is also listed on the website of the Verdaccio instance accessible at `http://localhost:4873/`. This, of course, is mostly useful for testing out a publishing process or for speeding up npm installations with a local cache. Most of the time, having a local npm registry is not really necessary.

One question might come up at this point: how can we make sure that a published package can be used by most users? What requirements need to be fulfilled for actually using a package in a client-based application running in the browser, as well as in a server-based application running in Node.js?

The concept of being pretty much target-independent is called being isomorphic. The terminology itself does not go uncriticized and some people actually prefer to call it universal. Having isomorphic code is great for gaining flexibility. Let's see what is needed to deploy isomorphic packages.

Writing isomorphic libraries

The holy grail of web development is the ability to write code not solely for the frontend or the backend but for both parts. Many frameworks and tools try to give us this capability.

To be accessible to multiple platforms, we not only need to ship multiple variants of our code but also only use APIs that are available on all supported platforms. For instance, if you want to make an HTTP request, then using `fetch` would be the right call for modern browsers. However, `fetch` was not available in less recent versions of Node.js. Therefore, you might need to solve this differently.

In the case of HTTP requests, there are already isomorphic libraries available – that is, libraries that will just do the right thing depending on the target runtime. You should only depend on these libraries.

> **Isomorphic fetch**
>
> The HTTP request problem can be solved in many ways – that is, by choosing an isomorphic library such as `axios` or `isomorphic-fetch`, the issue can be delegated to a dependency. The advantage of this method is that we do not need to find out what ways we need to follow on each platform. Additionally, testing and verification are much simpler that way.

For now, we will focus on providing multiple variants. If we want to publish our library with support for multiple module formats – say CommonJS and ESM – we can do that by extending the `package.json`. Setting `type` to `module` will tell Node.js that the module referenced by the `main` field actually follows ESM. In addition, we can define all of the package's exports explicitly – with an additional option to define what module to use depending on the used target platform and module system.

Let's see an example of this kind of configuration:

package.json

```
{
  // ... like beforehand
  "type": "module",
  "main": "dist/index.js",
  "exports": {
    ".": {
      "browser": {
        "require": "./lib/index.min.js",
        "default": "./dist/index.min.js"
      },
      "default": {
        "require": "./lib/index.js",
        "default": "./dist/index.js"
      }
    }
  }
}
```

In the case of our small library, there is a significant difference between the browser version and the non-browser version. However, for optimization, we've used minified modules for the browser, while all other platforms including Node.js will resolve to non-minified modules.

To create output suitable for CommonJS, we can use the `build` script that we've derived already:

```
$ esbuild src/*.js --platform=node --outdir=lib --format=cjs
```

The output for ESM is similar, but contains one important change:

```
$ esbuild src/*.js --platform=node --outdir=dist --format=esm
--define:__dirname="'.'"
```

The crucial change is to avoid using the `__dirname` global variable, which only works in Node.js using CommonJS. Instead, we just use the current directory. The change is not perfect, but should get the job done.

Right now, everything seems to be well prepared – but actually, it's not. The most important thing is still missing – the removal of the Node.js inbuilt package references. Our simple library references `fs` and `path`, but these packages do not exist in the browser. They would not know how to work there. Luckily, in this case, we have multiple solutions. The best one is arguably to replace the dynamic file read with a static import of the package's `package.json`:

index.js

```
import { name, version } from '../package.json';
export function getLibName() {
  return `${name}@${version}`;
}
```

Of course, this kind of algorithmic change is not always possible. In the given scenario, we also benefit from `esbuild`'s bundle option, which will include the necessary parts from the referenced JSON file to produce an output file that matches our expectations.

With these changes in mind, let's see how the `build` scripts are defined:

```
{
  // ... like beforehand
  "scripts": {
    "build-cjs-node": "esbuild src/*.js --platform=node
      --outdir=lib --format=cjs",
    "build-cjs-browser": "esbuild src/*.js --platform=node
      --outdir=lib --bundle --format=cjs --minify --entry-
      names=[name].min",
    "build-cjs": "npm run build-cjs-node && npm run build-
      cjs-browser",
```

```
    "build-esm-node": "esbuild src/*.js --platform=node
      --outdir=dist --format=esm",
    "build-esm-browser": "esbuild src/*.js --platform=node
      --outdir=dist --bundle --format=esm --minify --entry-
      names=[name].min",
    "build-esm": "npm run build-esm-node && npm run build-
      esm-browser",
    "build": "npm run build-cjs && npm run build-esm"
  }
}
```

It makes sense to define the scripts so that they can be run independently but also conveniently together without much effort. In many cases, the tool you've chosen has to be configured extensively to have the desired behavior. In the case of our example, `esbuild` was already quite equipped for the task – everything that we needed could be done via the command-line options.

One additional case that can be covered with an npm package is to actually provide a tool. Ideally, these are tools to be run with Node.js making it a cross-platform tool. Let's see how we can write and publish this kind of tool.

Publishing a cross-platform tool

Node.js would not be so powerful without its ecosystem. As we learned in *Chapter 1*, *Learning the Internals of Node.js*, relying on the power of its ecosystem was an elementary design decision. Here, npm takes the leading role by defining the package metadata in `package.json`, as well as the installation of packages.

During the installation of a package, a couple of things are happening. After the package has been downloaded, it will be copied to a target directory. For a local installation with npm, this is the `node_modules` folder. For a global installation with npm, the target will be globally available in your home directory. There is, however, one more thing to do. If the package contains a tool, then a reference to the tool will be put into a special directory, which is `node_modules/.bin` for a local installation.

If you go back to the code from the previous chapter, you will see that, for example, `jest` is available in `node_modules/.bin`. This is the same `jest` executable that we started with npx. Let's take the following:

```
$ ./node_modules/.bin/jest --help
```

We can compare it to this:

```
$ npx jest --help
```

Both will yield the same result. The reason is that npx for local installation is just a convenient tool to avoid writing out the path. As a reminder, you should opt for local installations over global installations.

> **npx and npm**
>
> npx is another command that comes together with the installation of npm. From a command perspective, npm is used to manage the dependencies, while npx is used to run packages. The npm utility also has a run subcommand, which runs commands that are defined in the scripts section of package.json, whereas npx runs commands as defined in the bin section of npm packages.

Now, the question is how can we create a package that also adds a script to the .bin folder so that it just works when installed? The answer lies in the package.json of our previous library.

Let's modify package.json a bit:

package.json

```json
{
  "name": "@foo/tool",
  "version": "1.0.0",
  "description": "A simple tool greeting the user.",
  "bin": {
    "hello": "lib/hello.js"
  },
  "license": "MIT"
}
```

We added a bin section that defines a single script to be referenced from the .bin directory. The reference should be called hello and pointed to the lib/hello.js file within this package.

Let's also add the script to run when hello is called:

hello.js

```js
#!/usr/bin/env node

// check that at least one argument has been provided
if (process.argv.length < 3) {
  console.log("No argument provided.");
  return process.exit(1);
```

```
}

// take the last argument
const name = process.argv.pop();
console.log(`Hello ${name}!`);
```

This will essentially check whether at least one argument was given and print a message in the console using the last argument.

Let's see the behavior when running directly via node:

```
$ node hello.js
No argument provided.
$ node index.js foo
Hello foo!
```

Now, the package can be published as before – for example, by choosing our local Verdaccio instance:

```
$ npm publish --registry http://localhost:4873
```

In a new project, you can now install the dependency and run the tool:

```
$ npm install @foo/tool --registry http://localhost:4873
$ npx hello bar
Hello bar!
```

With that, we have seen the most crucial aspects regarding the publishing process of npm packages. Let's recap what we have learned.

Summary

In this chapter, you have learned about what it takes to publish a package to an npm registry – whether it is an official or private one. You also touched on a commonly used npm registry in the form of Verdaccio.

Equipped with the knowledge from this chapter, you should now be able to write reusable libraries that work in browser-based applications as well as in Node.js-based applications. You are also now capable of publishing tools that are based on Node.js. In a sense, these tools are just libraries with some additional fields in their associated package metadata.

In the next chapter, we will have a look at a different approach to structuring code – placing multiple packages in a single repository known as a monorepo.

Structuring Code in Monorepos

In the previous chapter, you learned about everything to create and publish great libraries and tools to enhance your projects. While some packages are created in a bit of vacuum, most already have a consuming application in mind. In this case, having two separate repositories – that is, one for the application and one for the library – is quite some overhead. After all, any change to the library should be at least partially tested before the library is published. A good way to make this relation more efficient is to structure this code in a monorepo.

A **monorepo** is a single code repository that hosts multiple projects. Since we focus on Node.js projects, we can say that a monorepo is a repository containing multiple packages identified by their own package.json.

Today, monorepos are frequently used to power some of the largest Node.js project code bases in the world. If you want to properly read and contribute to projects such as Angular, React, or Vue, you'll need extensive knowledge about monorepos and the tools that make monorepos possible. For your own projects, a good structure – quite often provided by implementing monorepos – can also be crucial.

We will cover the following key topics in this chapter:

- Understanding monorepos
- Using workspaces to implement monorepos
- Working with Lerna to manage monorepos
- Working with Rush for larger repositories
- Integrating Turborepo instead of or with Lerna
- Managing a monorepo with Nx to enhance Lerna

Technical requirements

The complete source code for this chapter is available at https://github.com/PacktPublishing/
Modern-Frontend-Development-with-Node.js/tree/main/Chapter09.

The CiA videos for this chapter can be accessed at https://bit.ly/3EjGZTL.

Understanding monorepos

The structure of a dedicated repository has always been very similar; we have a single package.
json in the root, a single node_modules folder containing the resolved dependencies, and a set
of source and configuration files, usually scattered between the root and some specific folders such
as src. A quite popular setup is shown in *Figure 9.1*:

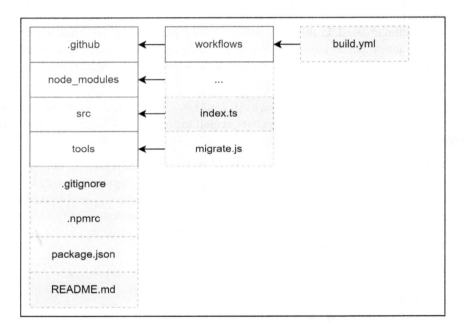

Figure 9.1 – Common setup for a repository with a single package

In the common setup, we have some folders for CI/CD pipeline definitions and potential tools that are useful for managing the repository, as well as auxiliary files such as project documentation. Of course, for a Node.js project, we'll see a node_modules directory, as well as a package.json file.

In contrast, a monorepo will contain multiple package.json files with multiple node_modules (or alternative) folders. Likewise, the source files and potentially some of the configuration will also be scattered across multiple locations. A very common structure is shown in *Figure 9.2* for the main part and *Figure 9.3* for the individual packages:

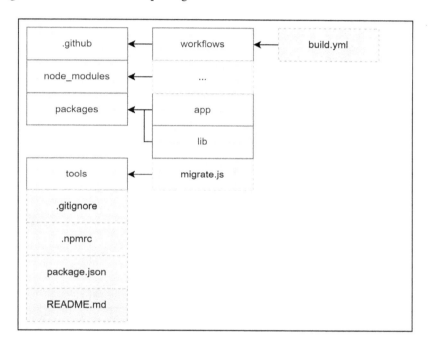

Figure 9.2 – Common setup for a repository with multiple packages

In comparison to *Figure 9.1*, the hierarchy of the outlined folders is a bit more sophisticated. Now, we don't see the source files immediately and need to descend into some of the directories inside the packages folder:

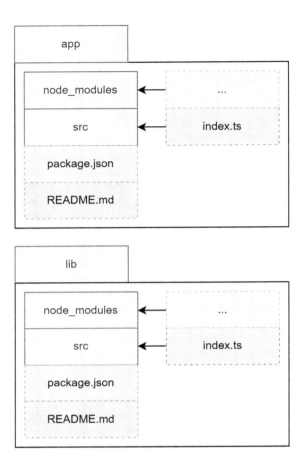

Figure 9.3 – The contents of the individual package directories

Ideally, the packages contained in a monorepo are crafted in such a way that makes them rather easy to extract later on. Let's say you have a specific library in your monorepo that should now be handled by another team. If your monorepo was created to serve as the single point of development for your current team, then transferring this library makes sense.

Quite often, common development concerns, such as the usual packages contained in the devDependencies of a package.json file, are concentrated in a dedicated package.json file. In many monorepos, this package.json file is found in the root directory of the monorepo. While this pattern makes sense from a maintenance point of view, it can also bring up challenges when it comes to library extraction. After all, you'll now need to decide what dependencies to add to restore the development capability of the extracted library.

In general, multiple challenges make supporting monorepos a task of its own. Here are some of the most pressing concerns:

1. How are dependencies efficiently shared to avoid installing the same dependencies over and over again?

2. How can packages be treated as if they are dependencies installed from a registry?

3. How can common tasks such as build steps be run in a way that works consistently?

Let's go through these one by one. For (*1*), the idea is that monorepos can be more efficient than just having many different directories, where you'd need to run `npm install` in each of them. Running `npm install` in each directory would be a massive overhead, duplicating not only direct dependencies but also indirect ones – that is, dependencies of installed dependencies.

While (*1*) is only a performance (installation time and disk space) concern, the issue with (*2*) is developer convenience. The reason for having a monorepo in the first place is to have packages that depend on one another in close proximity. This way, a bug should be visible at development time, rather than later at integration time when a package has already been published. The usual mechanism of npm for this is to use the `npm link` command, which will make a local package globally available for referencing. There are, however, multiple downsides to this mechanism. Additionally, it is not very convenient to use this command for every package.

Finally, the dependencies between the packages in a monorepo require special attention when running commands. In terms of (*3*), tasks such as building the source code need to be performed in reverse reference order. This means, that in the case that package *A* depends on package *B*, the build process of package *B* needs to be done before package *A* is built. The reason is that through the dependency, the content of package *A* may only build successfully if the content of package *B* has been fully created – that is, the package has been built. Similar constraints arise for testing and when publishing a package.

With this in mind, let's start with one of the easiest options for implementing a monorepo: leveraging the workspaces feature that comes with the most popular npm clients.

Using workspaces to implement monorepos

As the need for monorepos grew, npm clients tried to help users by incorporating them. The first of the big three was *Yarn*. Already, with the first version of Yarn, a new concept called **Yarn workspaces** was introduced, which was represented by a special field called `workspaces` in `package.json`:

package.json

```
{
  "name": "monorepo-root",
  "private": true,
```

```
    "workspaces": [
      "packages/*"
    ]
}
```

Yarn workspaces require a `package.json` at the root directory of the monorepo. This `package.json` won't be used for publishing and needs to have the `private` field set to `true`. The `workspaces` field itself is an array that contains the paths to the different packages. Wildcards using the `*` or `**` symbols – as shown here – are allowed.

With npm *v7*, the standard npm client also received a workspaces feature. The feature is pretty much the same as the implementation in Yarn. Here, we need to have a `package.json` in the root, too. Likewise, the behavior is controlled by a `workspaces` field.

Finally, the implementation in *pnpm* is a bit different. Here, we need a dedicated file called pnpm-workspace.yaml. This file contains the paths to the different packages:

pnpm-workspace.yaml

```
packages:
  - 'packages/*'
```

In contrast to the other two npm clients, with pnpm, you don't need a `package.json` file in the root directory. Since the workspaces definition is in a separate file, this file alone is sufficient to enable the workspaces feature of pnpm.

To illustrate that, let's create a new directory and add the preceding pnpm-workspace.yaml file to it. Then, create a `packages` subfolder. In there, add two more folders, p1 and p2. In each of these directories, run `npm init -y`. You can now modify the contained `package.json` files, adding some dependencies to both.

From the root directory with the pnpm-workspace.yaml file, run the following:

```
$ pnpm install
Scope: all 2 workspace projects
Packages: +5
+++++
Packages are hard linked from the content-addressable store to
the virtual store.
  Content-addressable store is at: /home/node/.local/share/
pnpm/store/v3
```

```
    Virtual store is at:                  node_modules/.pnpm
  Progress: resolved 5, reused 5, downloaded 0, added 5, done
```

While editing the respective `package.json` file is always possible, pnpm also makes it easy to add a dependency to some contained package – or workspace in the terminology of pnpm.

Let's say you want to add `react-dom` to the `p1` workspace:

```
$ pnpm add react-dom --filter p1
No projects matched the filters "/home/node/ Chapter09/
example01" in "/home/node/Chapter09/example01"
.                                              |   +2 +
Progress: resolved 5, reused 5, downloaded 0, added 0, done
```

The `--filter` argument allows you to select the workspaces where the dependency should be added. While full names are accepted, the names can also be specified with wildcards (`*`).

Specifying dependencies in monorepos

Dependencies on other packages contained in the same monorepo are declared just like any other dependency – in the corresponding `package.json` fields, such as `dependencies` or `devDependencies`. The specified version, however, is crucial here. You need to make sure to either match the version of the referenced package (e.g., `1.2.3` or `^1.0.0` would both correctly match a package in version `1.2.3`) or use the wildcard specifier `*`. Today, most package managers also support the special workspace protocol. With this, you can write `workspace:*` instead of a version to link against a package in another workspace.

The workspaces option is certainly appealing to optimize packages and make their linking quite easy; however, it fails to make common monorepo tasks more approachable or convenient. An alternative is to use a tool such as **Lerna** on top of a workspace.

Working with Lerna to manage monorepos

Lerna is one of the oldest tools for managing monorepos. We can even say to some degree that Lerna not only made monorepos manageable but also popular. Lerna is the backbone of some of the most important monorepos, such as Jest. It also was the original choice for projects such as *Babel* or *React*.

Originally, Lerna was mainly picked because it correctly installed and resolved all the packages. At this time, no package manager was capable of doing that intrinsically. However, today, Lerna is most often used together with the workspace features offered by the different package managers. Of course, you can still use the original mode of Lerna, where plain npm is used to install and link the different packages. So, how does Lerna fit into this new role when the whole installation is done by the chosen package manager anyway?

It turns out that Lerna is a really great task-running layer on top of a package manager. For instance, running a `package.json` script such as `build` in all the contained packages is as straightforward as invoking the following:

```
$ npx lerna run build
```

This would only run the script in the packages that contain this kind of script. In comparison, Yarn would actually error out if one of the packages did not have a `build` script.

To get started with Lerna, you need to initialize the current repository as a Lerna monorepo. For this, the `init` command can be used:

```
$ npx lerna init
```

Once initialized, the repository should contain a `lerna.json` and a `package.json` file. By inspecting these files, you'll notice that `lerna.json` contains a version (by default `0.0.0`), but `package.json` does not. This is intentional. Lerna will actually manage the version here. The default choice is uniform versioning – that is, all packages will always get the same version. The other option is independent versioning. Here, each package can have its own version number. This is handy if different packages have their own release cycle.

To enable independent versioning, we can change `lerna.json`:

---

lerna.json

```json
{
  // ... as beforehand
  "version": "independent"
}
```

Alternatively, we could also initialize the repository using the `--independent` flag for the `lerna init` command.

The `package.json` file contains the `workspaces` property. By default, this is configured to include all directories from the `package` directory as packages. In the given configuration, Lerna would use npm as a package manager. In any case, the whole package management is left to an actual npm client.

As mentioned, Lerna is really good at running tasks. What else is considered a strength of Lerna? The whole publishing and version management. We've already seen that Lerna knows two modes: independent and uniform versioning. In the independent versioning mode, Lerna will check the published versions with the current version that is about to be published. Only in the case of a new version will the `publish` command actually run.

Let's see how the packages from the previous example would actually be published with Lerna. We'll use a local registry running Verdaccio for this:

```
$ npx lerna publish --registry http://localhost:4873
lerna notice cli v5.5.2
lerna info versioning independent
lerna info Looking for changed packages since p1@1.0.1
? Select a new version for p1 (currently 0.0.0) Major (1.0.0)
? Select a new version for p2 (currently 0.0.0) Major (1.0.0)

Changes:
 - p1: 0.0.0 => 1.0.0
 - p2: 0.0.0 => 1.0.0

? Are you sure you want to publish these packages? Yes
lerna info execute Skipping releases
lerna info git Pushing tags...
lerna info publish Publishing packages to npm...
[...]
Successfully published:
 - p1@1.0.0
 - p2@1.0.0
lerna success published 2 packages
```

Without additional flags, Lerna will guide us through the whole publishing process. As we specified independent versioning, the tool will ask us the version to pick for each contained package. In the case here, we selected 1.0.0 for both packages.

Lerna also does a bit more than just running npm publish for each package. It is strongly related to **Git** as a version control system. It also ties the publish to the current commit and marks the publish via Git tags, which are automatically pushed to a potential origin such as **GitHub**.

Another thing that Lerna brings to the table is extensive information about a monorepo. Since Lerna needs to know which packages exist and what their relations are quite well, it also makes sense that this information is exposed to us.

A great command to use to see what exists in the current monorepo is lerna list:

```
$ npx lerna list --graph
lerna notice cli v5.5.2
```

```
lerna info versioning independent
{
  "p1": [
    "react",
    "react-dom"
  ],
  "p2": [
    "react",
    "react-dom"
  ]
}
lerna success found 2 packages
```

There are multiple options – all geared to fine-tune what information to include, exclude, and how to represent it. Ultimately, this is designed to make consumption in many ways possible. Independent of whether you consume this from a script or directly, the `lerna` tool has the right options to present the data accordingly.

Lerna has certainly been established as one of the go-to options for handling monorepos; however, its configuration options can be daunting, and making it efficient in a larger repository could be troublesome. An alternative is to use an opinionated tool instead. One of the best options in this category is **Rush**.

Working with Rush for larger repositories

While Lerna provided a lot of the utility that made monorepos possible at all, its configuration and flexibility also posed some challenges. Furthermore, finding best practices proved to be difficult. Consequently, plenty of quite opinionated alternatives to using Lerna have been born. One of the most successful ones is Rush from Microsoft.

Rush allows a variety of npm clients to be used. Classically, Rush used to be npm-only. Today, Rush recommends using pnpm, which is also the default client when setting up a monorepo with Rush.

To work efficiently with Rush, a global installation of the tool is recommended:

```
$ npm install -g @microsoft/rush
```

After a successful installation, the `rush` command-line utility can be used. As with `npm`, an `init` subcommand to actually initialize a new project exists:

```
$ rush init
```

This will create and update a couple of files. Most notably, you'll find a rush.json file in the current folder. This file needs to be edited next. However, before you continue, make sure to remove the files you don't need. For instance, Rush added a .travis.yml, which can be useful if you use Travis for your CI/CD pipelines. In case you don't know what Travis is or you know already that you don't want to use Travis, just delete that file.

Since with Rush, every package is added explicitly, there is no direct need for a packages subfolder. If you still prefer to group the contained packages in this way, you can of course do so.

In order to make Rush aware of the contained packages, we need to edit the rush.json file in the root folder. In our case, we want to add two new packages:

rush.json

```json
{
  // keep the rest as is
  "projects": [
    {
      "packageName": "p1",
      "projectFolder": "packages/p1"
    },
    {
      "packageName": "p2",
      "projectFolder": "packages/p2"
    }
  ]
}
```

Once the file is saved, you can run the following command – just make sure that the given directories really exist and contain a valid package.json file:

```
$ rush update
```

Among the given output, you should see some output containing messages similar to the ones we've seen when we introduced pnpm. As mentioned, under the hood, Rush uses pnpm to make package installation quite efficient.

Adding or updating a dependency in a package involves running rush add within the package directory. Let's say that we want to add react-router to p1:

```
$ cd packages/p1
$ rush add --package react-router
```

To run commands, Rush comes with two primitives. One is the generic `rushx` command, which can be seen as a wrapper around `npm run`. Let's say the `p1` package defines a `hello` command as follows:

packages/p1/package.json

```
{
  // as beforehand
  "scripts": {
    "hello": "echo 'Hi!'"
  }
}
```

Running this script can be done as follows:

```
$ cd packages/p1 && rushx hello
Found configuration in /home/node/examples/Chapter09/example02/
rush.json

Rush Multi-Project Build Tool 5.68.2 - Node.js 14.19.2 (LTS)
> "echo 'Hi!'"
Hi!
```

The other primitive is to use in-built commands such as `rush build` or `rush rebuild`. They assume that every package contains a `build` script. While the `rebuild` command will run all the `build` scripts, the `build` command actually uses a cache to enable an incremental build process – as in, reuse as much as possible from the output of the previous run.

While Rush is very restrictive and requires taking possession of the whole repository, an alternative is to use a more lightweight tool such as Turborepo.

Integrating Turborepo instead of or with Lerna

So far, we've seen quite a variety of tools in this chapter. While the workspaces feature of modern npm clients is already more than sufficient for smaller monorepos, larger ones require more dedicated tools to be manageable. In cases where Lerna is a bit too simplistic and Rush is too opinionated, another alternative exists – **Turborepo**, or Turbo for short. It can be seen as a replacement for or an addition to Lerna.

Starting from scratch is rather easy – Turbo comes with an npm initializer:

```
$ npm init turbo
```

This will open a command-line survey and scaffold the directory with some sample code. In the end, you should see a couple of new files being created, such as a `turbo.json` or a `package.json` file. Furthermore, Turbo creates `apps` and `packages` directories containing some sample code.

Let's show the strength of Turbo by running the `build` script:

```
$ npx turbo run build
```

In contrast to Lerna, this will not run the `build` script in each package – following the package graph. Instead, this will run one of the pipelines defined in `turbo.json`. In there, you can see the following:

turbo.json

```
{
  "$schema": "https://turborepo.org/schema.json",
  "pipeline": {
    "build": {
      "dependsOn": ["^build"],
      "outputs": ["dist/**", ".next/**"]
    },
    "lint": {
      "outputs": []
    },
    "dev": {
      "cache": false
    }
  }
}
```

The given `pipeline` property defines a set of Turbo `build` pipelines. Every given key (in the definition here, `build`, `lint`, and `dev`) can then be run via `turbo run`. The specifics of each pipeline are specified by its given value. For instance, the `dev` pipeline does not use a cache, while the `lint` pipeline does not produce any outputs. By default, each pipeline runs a script with the same name in each package.

The `build` pipeline here specifies some output directories that are cached to perform incremental builds. It also specifies that the `build` script has to be run in dependencies before it can run in the current package. Therefore, if you have two packages, `p1` and `p2`, where `p1` depends on `p2`, the build script of `p2` needs to run before the build script of `p1` can be invoked.

Besides the "in a different workspace" dependency (e.g., ^build), you can also specify "in the same workspace." For instance, if the build script depends on a prebuild script, you'd just write prebuild:

turbo.json

```
{
  "pipeline": {
    "build": {
      "dependsOn": ["^build", "prebuild"]
    }
  }
}
```

The turbo run command can also invoke multiple commands at the same time:

```
$ npx lerna turbo lint build
```

The result is a pretty efficient run since lint does not specify dependencies – so all linting can be done in parallel, while the build is executed hierarchically. The idea is illustrated in *Figure 9.4*:

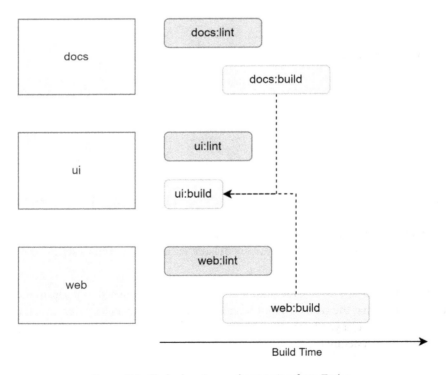

Figure 9.4 – Task planning and execution from Turbo

Turbo is not the only tool that can be used to make monorepos more efficient. A good alternative that goes beyond task running is Nx.

Managing a monorepo with Nx to enhance Lerna

Earlier in this chapter when we discussed Lerna, one thing we did not mention is that there is a special key in `lerna.json`, which is called `useNx` and configured to be `true`. This is a new addition to Lerna 5, which is now maintained by the people behind Nx – another popular solution for managing monorepos. So, what does this actually bring and how can it enhance Lerna – or any other monorepo management tool?

> **With Lerna or without?**
>
> Nx does not depend on Lerna and the use of Nx within Lerna is also optional. Therefore, the two technologies can be seen as non-exclusive – rather, they complete each other. In the end, it is your choice to decide which technologies you'd like to use. The example in this section, for instance, does not use Lerna.

We start with a new repository again. This time, we'll use the `nx-workspace` npm initializer provided by Nx:

```
$ npm init nx-workspace -- --preset=react
✓ Workspace name (e.g., org name)      · example05
✓ Application name                      · example
✓ Default stylesheet format            · css
✓ Enable distributed caching to make your CI faster · Yes
[...]
```

As with Turbo, we get a command-line survey. The initial preset (in this case, `react`) defines some of the questions that appear. There are other similarities to Turbo, too. For instance, running something is done via `nx`, such as the following:

```
$ npx nx build
```

This will look for the Nx `build` task executor of the current application (in this case, `example`) in a given environment (by default, `production`). Here is an explicitly written example:

```
$ npx nx run example:build:production
```

The task executor is specified in the `project.json` of a package. Nx uses plugins to actually run these executors; in the case of our sample project with the `react` preset, the `@nrwl/webpack` package is used as the plugin.

In order for Nx to work, each package requires either a `package.json` or a `project.json` file. Both can be specified, too. In this case, Nx will actually merge them internally to get the desired configuration. Usually, you'd want a `package.json` if you wanted to use npm scripts. The `project.json` file contains Nx task executors, which are a bit more powerful, but unfortunately, are beyond the scope of this quick introduction.

Let's stop here and recap what we learned in this chapter.

Summary

In this chapter, you learned how to organize multiple Node.js projects in a single repository known as a monorepo. You've seen different techniques and tools for maximizing efficiency and dealing with multiple packages and their dependencies.

You are now ready to deal with the largest code bases available. Independent of whether a code base just uses workspaces with one of the npm clients or some other tool such as Lerna on top of it, you are able to understand its structure, run commands, and add new packages in no time.

In the next chapter, we will conclude with a look at WebAssembly, which not only offers a lot of flexibility for code running in the browser but can also be used to run arbitrary languages in Node.js.

10
Integrating Native Code with WebAssembly

The whole point of actually using Node.js is convenience. Node.js never aspired to be the fastest runtime, the most complete one, or the most secure one. However, Node.js established a quick and powerful ecosystem that was capable of developing a set of tools and utilities to actually empower the web development standards that we are all used to today.

With the growth of Node.js, the demand for more specialized systems also increased. The rise of new runtimes that offered alternatives to Node.js actually resulted from this need. An interesting alternative can be found in the WebAssembly language. WebAssembly is a portable binary-code format like the **Java Virtual Machine (JVM)** or the **Microsoft Intermediate Language (MSIL)**. This makes it a potential compilation offering for any language – especially lower-level languages such as C or Rust.

In this chapter, you'll learn what WebAssembly has to offer, how you can integrate existing WebAssembly code in your Node.js applications, and how you can generate WebAssembly code yourself. By the end, you will be prepared to take your scripts to the next level – whether it is with WebAssembly itself or with WebAssembly running in Node.js.

We will cover the following key topics in this chapter:

- The advantages of using WebAssembly
- Running WebAssembly in Node.js
- Writing WASM with AssemblyScript

Technical requirements

The complete source code for this chapter is available at `https://github.com/PacktPublishing/Modern-Frontend-Development-with-Node.js/tree/main/Chapter10`.

The CiA videos for this chapter can be accessed at `https://bit.ly/3DPH53P`.

Advantages of using WebAssembly

WebAssembly (WASM) is a language without any runtime. Any kind of functionality – from allocating some memory to making an HTTP request – needs to be integrated by the consuming application. There are, however, some emerging standards such as the **WebAssembly System Interface (WASI)** that aim to bring a set of standard functionalities to any platform. This way, we can write platform-independent applications using WASM, with a runner integrating WASI.

> **WASI specification**
>
> The WASI specification covers everything that is needed to run WASM outside of a browser. Popular WASM runtimes such as **Wasmtime** or **Wasmer** implement WASI to actually run WASM applications. WASI specifies how system resources can be accessed by WASM. As a result, besides having WASI implemented in the runtime, the executed WASM code also needs to know (and use) the API provided by WASI. More details can be found at `https://wasi.dev/`.

Consequently, one of the advantages of WASM is its portability and ability to run in a sandbox. After all, there is no linking and ability to run system commands or access critical system resources.

Even something as simple as logging to the console (i.e., the equivalent of using `console.log()` in Node.js) needs to be provided by the WASI layer, which could leave the access decision for certain resources to the user.

Another advantage of WASM is that it is not a language directly. Therefore, we can actually use any language that supports WASM as a compilation target. As of today, most system languages such as C/C++, Rust, Zig, and Go support WASM generation. Finally, the "write once, run everywhere" principle of Java seems to be fulfilled.

Quite often, performance is considered another advantage of WASM. While WASM by itself can actually provide better performance than Node.js or similar runtimes, it will certainly still be slower than equivalent but very well-optimized native code. After all, this also just runs natively but with a bit less information, and in a more generic mode. Nevertheless, for some algorithms, the slowdown from WASM execution to native execution can be quite small, or even unnoticeable.

So, how is this all achieved? Well, first of all, the format of a WASM file is binary – that is, as efficient as possible. The structure in this binary is tailored to be parsed and executed really quickly. Instead of having high-level instructions such as loops, the language only offers labels and jump points – much like a true machine language.

In *Figure 10.1*, you can see the general flow and portability promise offered by WASM. As a developer, we only need to care about compiling to a `.wasm` file. If our tooling is capable of doing that, users are able to consume such files with the WASM runtime of their choice, which can be a browser or Node.js, but many other options exist too.

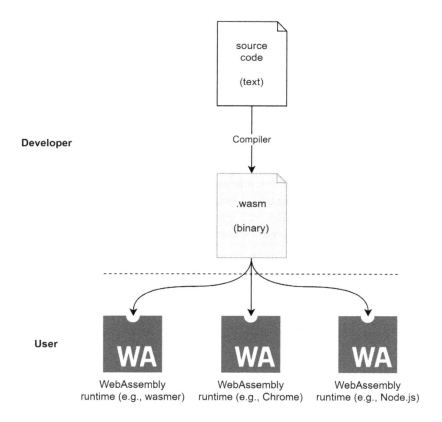

Figure 10.1 – Portability in WASM with a WASM binary

Like with machine languages, WASM has two formats – a text representation, which is great for seeing what's going on, and a corresponding binary representation. The text representation looks quite close to programming languages such as **Lisp**, with lower-level fragments resembling actual processor instructions.

Let's see an example of a WASM text representation for a library exporting a sum function to add two numbers together:

---

sum.wat

```
(module
(export "sum" (func $module/sum))
 (func $module/sum (param $0 i32) (param $1 i32)
   (result i32)
```

```
   local.get $0
   local.get $1
   i32.add
 )
)
```

Tools exist to translate the text representation into its binary counterpart. The most popular tool is **wat2wasm**, which has a powerful online demo, too. You can access it at `https://webassembly.github.io/wabt/demo/wat2wasm/`.

Adding the preceding example, you'll get a view as presented in *Figure 10.2*. You'll see that the online tool does a bit more than the text (upper left) to binary (upper right) translation. It also includes a small JavaScript playground (bottom left), which integrates the compiled WASM binary and runs it. The output of running code in the playground is then shown in the bottom-right corner.

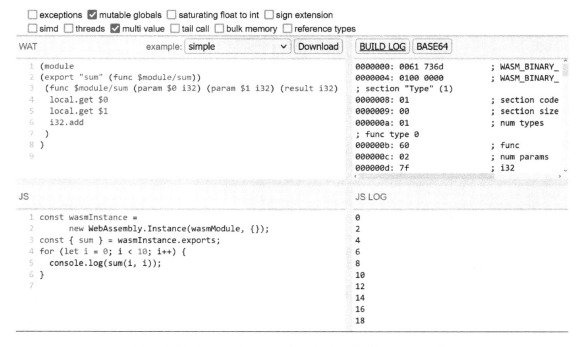

Figure 10.2 – The wat2wasm online tool applied to our example

Now that we know what WASM is, how it works, and what advantages it offers, it's time to see how we can run it and – of course – also integrate it with Node.js. This makes our scripts even more powerful than beforehand, allowing platform-independent, almost native code to be integrated in a reliable, well-performing, and secure way.

Running WASM in Node.js

Node.js has a direct integration of WASM via the `WASM` object. The API is exactly the same as in the browser, allowing us to potentially share the code between Node.js and browsers to integrate a compiled WASM file.

There are three functions in the API of `WASM`. We can compile an existing binary, transforming it into a WASM runtime module. This module can then be activated using the `instantiate` method. We can also validate an existing binary – to check whether a given file is indeed a valid WASM binary. All methods are asynchronous and return `Promise`.

Let's see an example using a WASM binary, `sum.wasm`, which exports a single function (`sum`) and adds two numbers together:

app.mjs

```
import{ readFile } from 'fs/promises';

const content = await readFile('./sum.wasm');
const wasm = await WebAssembly.compile(content);
const instance = await WebAssembly.instantiate(wasm);
const { sum } = instance.exports;
console.log(sum(2, 3)); // logs 5
```

Node.js makes the integration of WASM even more convenient by providing a `wasi` package out of the box. This package fulfills the WASI specification, allowing us to access system resources within WASM applications running in Node.js.

To see what the integration of a WASM module that depends on WASI looks like, we'll build a small application later, which will make use of WASI and be integrated into Node.js. The Node.js integration will look like this:

app.mjs

```
import { readFile } from "fs/promises";
import { WASI } from "wasi";
import { argv, env } from "process";

const wasi = new WASI({
  args: argv,
  env,
```

```
});

const api = { wasi_snapshot_preview1: wasi.wasiImport };
const path = "./echo.wasm";
const content = await readFile(path);
const wasm = await WebAssembly.compile(content);
const instance = await WebAssembly.instantiate(wasm, api);

wasi.start(instance);
```

At least with Node.js version 18, the `wasi` package is not active. To actually run the preceding application, you'll need to add the `--experimental-wasi-unstable-preview1` flag:

```
$ node --experimental-wasi-unstable-preview1 app.mjs
```

The specifics of running the preceding example are explored in the next section.

While running WASM in Node.js is great, we might also want to write some code ourselves. Of course, if you have any knowledge of languages such as C or Rust, you can use those with WASM as a compilation target. In general, however, for developers with a JavaScript background, a nice alternative exists with AssemblyScript.

Writing WASM with AssemblyScript

While there are many options to actually generate valid WASM, one of the most attractive ways is to use AssemblyScript. AssemblyScript is a language that looks and feels quite similar to TypeScript, making it rather easy to learn from a syntax perspective. Under the hood, however, there are still some concepts relating to WASM that need to be known in order to write mid-sized to larger AssemblyScript applications or libraries.

One of the core concepts of AssemblyScript is to model the different data types used in WASM. For instance, using integers requires the use of the `i32` type.

Let's have a look at some example code. We'll start with a small function that expects two parameters, adds them up, and returns the result:

module.ts

```
export function sum(a: i32, b: i32): i32 {
  return a + b;
}
```

With the exception of the i32 type, everything in the preceding example looks and feels just like TypeScript. Even the file extension indicates a TypeScript file.

To actually compile the preceding code to WASM, you'll need the assemblyscript package. Like typescript, you can either install this package globally or locally.

Once AssemblyScript is installed, you can run the asc utility to compile the source code to a valid WASM binary:

```
$ npx asc module.ts --outFile sum.wasm --optimize
```

AssemblyScript can also be very helpful to scaffold a project structure that works – not only to compile source code but also to run WASM in the browser. This provides a nice way of writing code that works on multiple platforms, including various operating systems, browsers, and devices:

```
$ npx asinit .
Version: 0.21.6
[...]
  ./assembly
  Directory holding the AssemblyScript sources being compiled
to WebAssembly.

  ./assembly/tsconfig.json
  TypeScript configuration inheriting recommended
AssemblyScript settings.

  ./assembly/index.ts
  Example entry file being compiled to WebAssembly to get you
[...]
  ./index.html
  Starter HTML file that loads the module in a browser.

The command will try to update existing files to match the
correct settings [...]
Do you want to proceed? [Y/n] Y
```

With the generated structure in place, we can go ahead and try to make our previous example work – for instance, in a web browser.

For this, modify index.ts in the assembly directory of the scaffolded project folder. Replace its content with the preceding snippet containing the sum function. Now, open index.html in the project's root. Change the import statement to obtain sum instead of add.

The script part of the `index.html` file should now look like this:

```
import { sum } from "./build/release.js";
document.body.innerText = sum(1, 2);
```

Now, you can build and run everything using the `asbuild` script that was added during the scaffolding process:

```
$ npm run asbuild
$ npm start
```

Now, a small web server should be running at port `3000`. Accessing `http://localhost:9000` brings you to an almost empty web page. What you should see is that **3** is written in the top-left corner of the page. This is the result of calling the exported `sum` function from our WASM library.

> **Debugging WASM**
>
> A WASM module can be debugged in the same way as any other web application. The browser offers a visual debugger that can be used for inspection. By using source maps for WASM, the original code can actually be debugged instead of the not-so-easily readable WASM. AssemblyScript is also capable of producing WASM source maps. Here, the source map destination file has to be specified after the `--sourceMap` CLI flag.

AssemblyScript can also be used to create WASM applications and libraries built on top of WASI. Let's see how that would work. We start with a new project, where we add `assemblyscript` as well as `as-wasi` as dependencies, followed by scaffolding a new `AssemblyScript` project:

```
$ npm init -y
$ npm install assemblyscript as-wasi --save-dev
$ npx asinit . -y
```

Now, we can modify the `assembly/index.ts` file with the following code, using the `wasi` package.

index.ts

```
import "wasi";
import { Console, CommandLine } from "as-wasi/assembly";

const args = CommandLine.all;
const user = args[args.length - 1];


Console.log(`Hello ${user}!`);
```

By importing the `wasi` package, the whole module gets transformed into a WASI-compatible entry point. This allows us to use the abstractions from the `as-wasi` package, such as `Console` to access the console or `CommandLine` to obtain the provided command-line arguments.

To build the code we invoke the `asc` utility with the following arguments:

```
$ npx asc assembly/index.ts -o echo.wasm --use abort=wasi_abort
--debug
```

This instructs AssemblyScript to build the application found in `assembly/index.ts`. The generated WASM will be stored in `echo.wasm`. Through the `--debug` flag, we instruct `asc` to create a debug build.

A debug build can be done very fast, as the compiler does not need to invest in any optimizations. Besides a faster compilation time, the absence of further optimizations also can give us better error messages for critical failures later at runtime.

Importantly, the binding for the `abort` command (usually taken from an implied `env` import to the WASM module) is set to use the `abort` method provided by WASI.

Now, we can add the Node.js module, `app.mjs`, using the `wasi` package from the previous section. Don't forget to add the necessary command-line argument. Since this will print a warning, we might want to add `--no-warnings` to suppress it:

```
$ node --experimental-wasi-unstable-preview1 --no-warnings app.
mjs Florian
Hello Florian!
```

Equipped with this knowledge, you can now go ahead and write simple programs compiling to WASM, too. Let's recap what you learned in this chapter.

Summary

In this chapter, you extended your knowledge of potential source code files running in Node.js. You are now familiar with running WASM – a lower-level portable binary-code language that can be used as a compilation target by many programming languages.

WASM can help you to write functionality once and run it on multiple platforms. Since WASM can be sandboxed very well, it is a good contender for the next wave of containerized computing, where performance and security are valued highly. You now know how to write WASM using AssemblyScript. You are also empowered to integrate created WASM modules in Node.js.

In the next and final chapter, we will take a look at the use of JavaScript beyond Node.js. We'll see that other runtimes exist, which are partially compatible with the Node.js ecosystem – providing a great drop-in replacement that can be handy for multiple use cases.

11

Using Alternative Runtimes

So far, you've seen what advantages and benefits the Node.js ecosystem offers to create great web applications. However, as with almost everything, there are a few downsides to the design decisions forming what we refer to as Node.js.

One of the biggest challenges in Node.js is the so-called *dependency hell* – where many small packages are put together to create a slightly larger package. Another challenge is that Node.js is not guarding any of these dependencies from accessing system resources. As such, importing anything from a third-party package may have unwanted side effects.

While ecosystem reliability and security can help us guard against dependency hell, improving performance is also an important strategy. Overall, the performance of Node.js can be regarded as decent; however, certain areas such as package resolution or processor core utilization could be improved by a fair share. Hence, performance is another area that could be regarded as a downside.

In this chapter, you'll get to know the two most popular alternative runtimes for mitigating some of the disadvantages that come with Node.js. To evaluate these alternatives in depth, we will keep a closer eye on their compatibility status with the existing Node.js ecosystem.

We will cover the following key topics in this chapter:

- Exploring the Deno runtime
- Using Bun for bundling web apps

Technical requirements

The complete source code for this chapter is available at https://github.com/PacktPublishing/Modern-Frontend-Development-with-Node.js/tree/main/Chapter11.

The CiA videos for this chapter can be accessed at https://bit.ly/3Uqi9aq.

Exploring the Deno runtime

While Node.js is a tremendous success story, not everyone is a fan. Some critics say that the huge fragmentation combined with the lack of system controls offers too great an attack surface. In the past, we've seen countless attacks that have abused the vulnerabilities introduced by exactly this problem.

Another issue is that Node.js did have to invent a lot of APIs – for example, to interact with the filesystem. There was no API available in the browser that looked similar to what was desired. Of course, as we now know, the browser APIs kept improving and even things such as filesystem access are implemented there. However, the APIs never aligned, mostly because the variants for Node.js are neither controllable nor fully asynchronous.

Surely, the aforementioned problems were all known for a while, but it took several years until an alternative implementation to solve these issues appeared. Again, it was Ryan Dahl – the original creator of Node.js – who worked on the solution. The solution is called **Deno**.

The main benefits of Deno are as follows:

- It introduces system access controls to allow or block access to resources such as the filesystem.
- It uses explicit imports instead of magically resolved ones – no more implied package lookups or index files.
- It tries to be interchangeable with the browser – bringing exclusively native browser APIs instead of custom ones.
- It features first-class TypeScript support, not only improving the development experience but also strengthening the reliability of written code.
- It comes with handy tooling, such as an application bundler out of the box – reducing the need to install dependencies for starting development.

Under the hood, Deno uses the Rust programming language instead of C++. Here, the choice was made to avoid any potential memory leaks or vulnerabilities that are just a bit more likely with C++ than **Rust**. This also means that *libuv*, which is the main driver for Node.js as discussed in *Chapter 1, Learning about the Internals of Node.js*, is gone. Instead, Deno uses another event loop system called **Tokio**. Still, both runtimes use **V8** to actually run JavaScript.

> Tokio
>
> Tokio is an asynchronous runtime for Rust applications providing everything needed for interacting with networks. It is reliable, fast, and flexible. Besides being Rust-native, one of the core reasons for Deno to use Tokio was that it is also quite easy to integrate. Tokio comes with I/O helpers, timers, filesystem access, synchronization, and scheduling capabilities, making it a complete libuv replacement. More information can be found at `https://tokio.rs/`.

The architecture of Deno is shown in *Figure 11.1*. Notably, the diagram is almost an exact match for *Figure 1.1*, which showed the architecture of Node.js. The most striking difference is the acceptance of TypeScript, which will be translated into JavaScript by a combination of swc (transpilation) and tsc (type checking). Another crucial difference is the additional isolation layer:

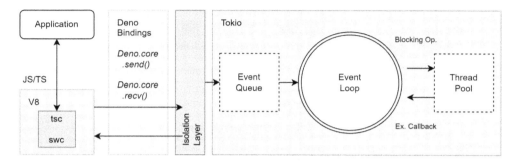

Figure 11.1 – The architecture of Deno

The installation of Deno can be done on the command line. For instance, on macOS and Linux, you can just run the following Shell script:

```
$ curl -fsSL https://deno.land/x/install/install.sh | sh
```

Whereas on Windows, you can use the PowerShell for this:

```
$ irm https://deno.land/install.ps1 | iex
```

Alternative installations for common application package managers such as Scoop, Chocolatey, or Homebrew exist, too.

To try Deno, you can run the following script:

```
$ deno run https://deno.land/std/examples/welcome.ts
Download ⬚ https://deno.land/std/examples/welcome.ts
Warning Implicitly using latest version (0.159.0) for https://
deno.land/std/examples/welcome.ts
Welcome to Deno!
```

There are a couple of things happening already. First, we are not using a local source to run, but an address. Second, since this is an address, the source needs to be downloaded. Third, Deno always prefers to receive explicit versions, so it will complain that we just used whatever version of stdlib here. Instead, it redirects to the latest version, which was 0.159.0 at the time of writing.

Finally, if you run the script again, you'll just see the output without any download or warning. This is due to Deno's cache. In order to stay well performing, every downloaded module is assumed to be immutable and will be cached locally. Future references will therefore not require another download, which makes their startup time acceptable.

The big question is now: can Deno also just run Node.js libraries and applications? The unsatisfying answer is maybe. In theory, just JavaScript files can be used – however, Deno only supports ESM modules. Since many Node.js libraries are written using CommonJS, we would not have any success here.

As mitigation, we could just transpile a package – bundle it into one file and run it without any trouble – but even then, we might face the issue of incompatibility with the ecosystem, as standard packages such as `fs` are available in Node.js but not in Deno.

A better way out of this is to use the *Node compatibility mode* of Deno. Before version *1.25*, it worked by running `deno` with the `--unstable` and `--compat` flags. Right now, Deno seems to only allow this via custom imports. Let's try this out to see it in action. For this, you can create a new Node.js project with a single third-party package and some code using it:

```
$ npm init -y
$ npm install axios --save
```

To test this, the following code provides a solid basis:

---

index.node.mjs

```
import axios from 'axios';
import { writeFile } from 'fs/promises';

const { data } = await
  axios.get('https://jsonplaceholder.typicode.com/photos');
const thumbnails = data.map(item => item.thumbnailUrl);
const content = JSON.stringify(thumbnails, undefined, 2);

await writeFile('thumbnails.json', content, 'utf8');
```

The code uses a third-party dependency made for Node.js together with a Node.js core module. It also makes use of modern features such as top-level `await` statements.

You can try running this with Node.js to see it working, but more interesting is the case of running this with Deno:

```
$ deno run index.node.mjs
error: Relative import path "axios" not prefixed with / or ./
```

```
or ../
    at file:///home/node/examples/example01/index.mjs:1:19
```

As mentioned, by default, Deno requires explicit paths. Without them, Deno does not work. Let's modify this code to reflect the compatibility:

index.deno.mjs

```
import axios from 'npm:axios';
import { writeFile } from
  'https://deno.land/std@0.159.0/node/fs/promises.ts';

const { data } = await
  axios.get('https://jsonplaceholder.typicode.com/photos');
const thumbnails = data.map(item => item.thumbnailUrl);
const content = JSON.stringify(thumbnails, undefined, 2);

await writeFile('thumbnails.json', content, 'utf8');
```

While the majority of the preceding code remains unchanged in comparison to `index.node.mjs`, the imports have been adapted slightly. The referenced npm packages need to be referenced using the npm: protocol. For Node.js core modules, we can refer to the `std/node` modules provided by Deno.

Now, we can run the code with the –unstable flag:

```
$ deno run --unstable index.deno.mjs
✓ Granted env access to "npm_config_no_proxy".
✓ Granted env access to "NPM_CONFIG_NO_PROXY".
✓ Granted env access to "no_proxy".
✓ Granted env access to "NO_PROXY".
✓ Granted env access to "npm_config_https_proxy".
✓ Granted env access to "NPM_CONFIG_HTTPS_PROXY".
✓ Granted env access to "https_proxy".
✓ Granted env access to "HTTPS_PROXY".
✓ Granted env access to "npm_config_proxy".
✓ Granted env access to "NPM_CONFIG_PROXY".
✓ Granted env access to "all_proxy".
✓ Granted env access to "ALL_PROXY".
```

☑ Granted read access to "/home/rapplf/.cache/deno/npm/node_modules".

☑ Granted read access to "/home/rapplf/.cache/deno/node_modules".

☑ Granted read access to "/home/rapplf/.cache/node_modules".

☑ Granted read access to "/home/rapplf/node_modules".

☑ Granted read access to "/home/node_modules".

☑ Granted read access to "/node_modules".

☑ Granted net access to "jsonplaceholder.typicode.com".

☑ Granted write access to "thumbnails.json".

As we did not provide any additional CLI flags, Deno will run in a mode where every resource request will be reflected by a question on the command line. In the session here, every request was confirmed with *yes*, granting the access request.

Alternatively, we could have used a Deno feature that we discussed already in *Chapter 2, Dividing Code into Modules and Packages*, while discussing import maps. Let's try to run our *unmodified* file again with the following import map:

importmap.json

```
{
  "imports": {
    "axios": "npm:axios",
    "fs/promises":
      "https://deno.land/std@0.159.0/node/fs/promises.ts"
  }
}
```

The job of the import map is to teach Deno what to look for. Originally, Deno could not make sense of an import to `axios`, but now it knows that this should be resolved via npm. Similarly, the core Node.js packages can be added in there, too.

This time, we set the `--allow-all` flag to skip all the access confirmations:

```
$ deno run --unstable --import-map=importmap.json --allow-all
index.node.mjs
```

And… it just works. No more work needed – all done with Deno primitives. Of course, quite often full compatibility cannot be achieved so easily.

While Deno is mostly focused on security, a presumably even more interesting area is performance. This is where another alternative shines, which is called Bun.

Using Bun for bundling web apps

While Deno seems quite different from Node.js on first glance, it also offers a lot of similarities. After all, both runtimes use V8 and can work with ESMs, but what if you want to be even more compatible with Node.js? Another approach is to be Node.js-compatible without using libuv or V8 at all. Enter **Bun**.

Bun is an alternative to Node.js that follows the approach of Deno in terms of developer friendliness. Here, tooling such as a npm client or an application bundler is also included out of the box. However, to speed things up significantly, Bun does not use libuv and V8. Instead, Bun is created using the programming language **Zig** and uses **JavaScriptCore** as its JavaScript runtime. JavaScriptCore is also the runtime behind the **Webkit** browser engine, empowering browsers such as **Safari**.

The main benefits of Bun are as follows:

- It comes with useful utilities out of the box, such as a bundler, a transpiler, a package manager, and a task runner.
- It outperforms Node.js, especially in terms of startup performance or request handling.
- It embraces the Node.js ecosystem, but also includes some standard web APIs such as `fetch` or `WebSocket`.

A comparison of the high-level architecture of Node.js and Bun is shown in *Figure 11.2*. Most importantly, while extra tools such as a package manager or a bundler are required with Node.js, Bun comes with batteries already included. All these tools are available after installation – and since all of these tools are integrated into the Bun executable, they provide the best performance possible:

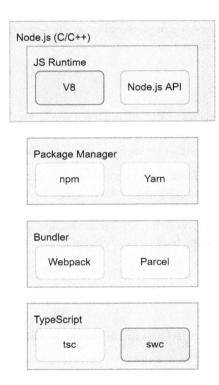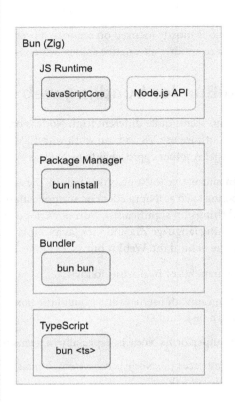

Figure 11.2 – High-level comparison of Node.js and Bun

As with Deno, Bun can be installed via a Shell script. At the time of writing, Bun is not available as a direct installation for Windows. Instead, you'll need to fall back to the **Windows Subsystem for Linux** (**WSL**) if you want to try out Bun.

To install Bun on macOS and Linux, you can run the following Shell script:

```
$ curl https://bun.sh/install | bash
```

Running a simple example (hello.ts) with Bun looks as follows:

```
$ bun run hello.ts
Hello from Bun!
```

In the preceding example, the code is really simple – just using the console output here:

hello.ts

```
console.log('Hello from Bun!');
```

An interesting aspect of Bun is that it also has the ability to automatically create a server. If we use a default export with a fetch function, then Bun will create a server, which, by default, runs on port 3000. The port can also be changed by having another property called port in there:

http.ts

```
export default {
  fetch() {
    return new Response("Hello from Bun!");
  },
};
```

Calling bun run http.ts will open the server. To see the result, go to the http:// localhost:3000 address using your browser.

Finally, let's use Bun as a bundler for the small demo project we did in *Chapter 6*. The first thing you should notice is that you don't need any development dependencies – just the runtime ones. Also, instead of running npm install or similar, you should resolve the dependencies via bun install:

```
$ bun install
bun install v0.1.13
  + react@18.2.0
  + react-dom@18.2.0
  + react-router-dom@6.4.2
  + video.js@7.21.0

  32 packages installed [2.21s]
```

Frankly, react, react-dom, react-router-dom, and video.js comprise only four packages, but their installation speed is still quite good. Now, it's time to bundle the JavaScript code:

```
$ bun bun src/script.tsx
[...]
  2.34 MB JavaScript
        58 modules
```

```
    20 packages
107.61k LOC parsed
    62ms elapsed
Saved to ./node_modules.bun
```

The result is quite different to the bundlers we've seen beforehand. We get a single file, node_modules. bun, which contains the resulting JavaScript, as well as all the associated metadata. The file itself is an executable – ready to spit out the contained code.

Extracting the JavaScript contained in the node_modules.bun file can be done by running the executable – and piping the output to a JavaScript file. For instance, see the following:

```
$ ./node_modules.bun > dist/app.js
```

Is this sufficient for all our bundling needs? Certainly not. Right now, the integrated bundler is essentially ignoring our code and only bundling together the code from the external packages sitting in the node_modules directory. However, even if our code was bundled, the process is not really ideal. Currently, Bun only considers JavaScript, TypeScript, JSON, and CSS files. There is no way to include assets such as images or videos.

For the future, all these capabilities are planned. While Bun (in version *0.1.13*) is still experimental technology, what is out there is promising. All things considered, it's certainly something to keep on the radar, but nothing that can be actively used to create production-ready code.

Let's recap what you've learned in this chapter.

Summary

In this chapter, you learned why alternatives to Node.js exists and what the most popular options are. You've explored what Deno is all about and how it distinguishes itself from Node.js. You've also seen an up-and-coming alternative with Bun.

Equipped with this knowledge, you are not only capable of writing tools that might be able to run in other runtimes than Node.js but you are also capable of deciding where your existing tools should run. Overall, this does not constrain you to the disadvantages of Node.js and gives you freedom to make the right choice aligned with the problem you want to solve.

Epilogue

In general, it makes sense to view Node.js as a great helper for getting the job done. The whole ecosystem – from its module system to its command-line utilities and from its libraries to its frameworks – is vast. Almost every problem has been cracked and a solution has been published.

I hope that with this book, you have a proper guide to walk you through the jungle of available helpers, making you not only a more efficient user of Node.js but also a contributor. While the existing tools are all helpful and powerful, they are certainly not the end of the line. Everyone has a unique view and things progress all the time. Don't wait for somebody else to solve a problem – tackle it yourself and share your solution.

All the best!

Index

Other Books You May Enjoy

If you enjoyed this book, you may be interested in these other books by Packt:

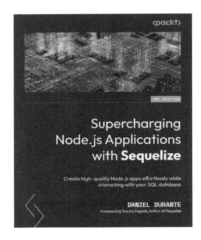

Supercharging Node.js Applications with Sequelize

Daniel Durante

ISBN: 978-1-80181-155-2

- Configure and optimize Sequelize for your application
- Validate your database and hydrate it with data
- Add life cycle events (or hooks) to your application for business logic
- Organize and ensure the integrity of your data even on preexisting databases
- Scaffold a database using built-in Sequelize features and tools
- Discover industry-based best practices, tips, and techniques to simplify your application development

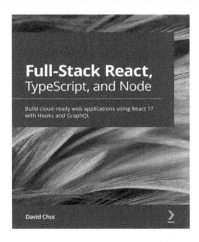

Full-Stack React, TypeScript, and Node

David Choi

ISBN: 978-1-83921-993-1

- Discover TypeScript's most important features and how they can be used to improve code quality and maintainability
- Understand what React Hooks are and how to build React apps using them
- Implement state management for your React app using Redux
- Set up an Express project with TypeScript and GraphQL from scratch
- Build a fully functional online forum app using React and GraphQL
- Add authentication to your web app using Redis
- Save and retrieve data from a Postgres database using TypeORM
- Configure NGINX on the AWS cloud to deploy and serve your apps

Packt is searching for authors like you

If you're interested in becoming an author for Packt, please visit `authors.packtpub.com` and apply today. We have worked with thousands of developers and tech professionals, just like you, to help them share their insight with the global tech community. You can make a general application, apply for a specific hot topic that we are recruiting an author for, or submit your own idea.

Share your thoughts

Now you've finished *Modern Frontend Development with Node.js*, we'd love to hear your thoughts! Scan the QR code below to go straight to the Amazon review page for this book and share your feedback or leave a review on the site that you purchased it from.

`https://packt.link/r/1-804-61829-2`

Your review is important to us and the tech community and will help us make sure we're delivering excellent quality content.

Download a free PDF copy of this book

Thanks for purchasing this book!

Do you like to read on the go but are unable to carry your print books everywhere?

Is your eBook purchase not compatible with the device of your choice?

Don't worry, now with every Packt book you get a DRM-free PDF version of that book at no cost.

Read anywhere, any place, on any device. Search, copy, and paste code from your favorite technical books directly into your application.

The perks don't stop there, you can get exclusive access to discounts, newsletters, and great free content in your inbox daily!

Follow these simple steps to get the benefits:

1. Scan the QR code or visit the link below:

https://packt.link/free-ebook/9781804618295

2. Submit your proof of purchase
3. That's it! We'll send your free PDF and other benefits to your email directly

www.ingramcontent.com/pod-product-compliance
Lightning Source LLC
Chambersburg PA
CBHW060559060326
40690CB00017B/3756